Cambridge Studies in Historical Geography

AGRICULTURAL REVOLUTION IN ENGLAND

Cambridge Studies in Historical Geography 23

Series editors:
ALAN R. H. BAKER, RICHARD DENNIS, DERYCK HOLDSWORTH

Cambridge Studies in Historical Geography encourages exploration of the philosophies, methodologies and techniques of historical geography and publishes the results of new research within all branches of the subject. It endeavours to secure the marriage of traditional scholarship with innovative approaches to problems and to sources, aiming in this way to provide a focus for the discipline and to contribute towards its development. The series is an international forum for publication in historical geography which also promotes contact with workers in cognate disciplines.

For a full list of titles in the series, please see end of book.

AGRICULTURAL REVOLUTION IN ENGLAND

The transformation of the agrarian economy 1500–1850

MARK OVERTON

Professor of Economic and Social History
University of Exeter

Published by the Press Syndicate of the University of Cambridge
The Pitt Building, Trumpington Street, Cambridge CB2 1RP
40 West 20th Street, New York, NY 10011-4211, USA
10 Stamford Road, Oakleigh, Melbourne 3166, Australia

First published 1996

Printed in Great Britain at the University Press, Cambridge

A catalogue record for this book is available from the British Library

Library of Congress cataloguing in publication data

Overton, Mark.
    Agricultural revolution in England: the transformation of the
agrarian economy 1500–1850 / Mark Overton.
        p.   cm. – (Cambridge studies in historical geography 23)
    Includes bibliographical references and index.
    ISBN 0 521 24682 2
    1. Agriculture – Economic aspects – England – History.   2. England –
– Economic conditions.   I. Title.   II. Series.
HD1930.E5093   1996
338.1'61'0942 – – dc20   95-33963 CIP

ISBN 0 521 24682 2 hardback
ISBN 0 521 56859 5 paperback

KS

Contents

Figures

Tables

Preface

Agrarian history is currently split into two camps: one primarily concerned with the activity of farming (sometimes referred to as 'cows and ploughs' agricultural history), and the other with a rural history that is more concerned with the wider social and cultural aspects of the countryside. Studies of the 'agricultural revolution' (and most of my own research) have fallen into the 'cows and ploughs' category and this has undoubtedly influenced the approach of this book. But despite these influences I have tried to extend from the *agricultural* ('the science and art of cultivating the soil') to the *agrarian* ('relating to the land'), in order to widen the terms of reference for considering the 'agricultural revolution'. This has meant that I have relied on the work of others for many parts of the book, and I am grateful to those historians whose ideas I have absorbed since the 1970s. They are too numerous to list; some of them appear in the bibliography, but there are many others.

The day after the text of the book was completed the editor of a well-known historical journal asked me to referee a manuscript for publication. It took the form that many articles do: an opening paragraph setting out the established view, followed by new ideas and evidence overturning that view. The opening paragraph of the article was almost identical to a paragraph in this book. This was a salutary reminder that our knowledge and understanding of the past is constantly changing, and, while I hope the main conclusions to this book may survive for a little longer than that paragraph, I also hope the flow of articles on agricultural and agrarian history may long continue.

I have not encumbered the text with footnotes, but most of my sources are listed in the *Bibliography*, and the *Guide to further reading* should cut a pathway into them. The sources of quotations are given, and, more importantly, the sources of the tables and figures, so that those who so wish can rework the original material.

This book has its origins in an undergraduate lecture course, which started in Cambridge and continued in Newcastle, and I am grateful to the successive generations of students for giving me the opportunity to develop my ideas and arguments. The SSRC (as it then was) funded me for a year during 1984–5 when research on the book began. I was also fortunate to hold a Visiting Fellowship at All Souls College, Oxford, during 1992–3, when much of the text was written. I must thank Olive Teesdale, Eric Quenett, Gary Haley, and especially Ann Rooke, for drawing the diagrams. I owe an enormous debt to my long-suffering publishers who have been waiting patiently for this book for over a decade.

Meemee has been forced to share me with this book for as long as she has known me. Her support has been invaluable, not only in reading the entire text several times and correcting the proofs, but in creating an environment which made work possible.

A note on weights, measures, money and boundaries

The weights and measures used in this book are the ones contemporaries used. The standard unit for grain was the bushel (roughly 36 litres) of 4 pecks or 8 gallons, and 8 bushels made a quarter (about 2.9 hectolitres). The weight of a bushel of grain can vary, but a bushel of wheat weighs about 56 pounds or roughly 25 kilograms, a bushel of barley 48 pounds, and a bushel of oats 38 pounds. The unit of measurement for area was the acre, equivalent to about 0.4 hectares; linear measurement was by the mile of about 1.6 kilometers. When area, weight and volume are related in measures of yield, 20 bushels per acre (the average wheat yield in England around 1800) is roughly equivalent to half a ton per acre, which is about 500 kilograms per acre, 1.2 tonnes per hectare, and 17.4 hectolitres per hectare.

Before 1971 the English pound (£) consisted of 20 shillings (s.), each shilling comprised 12 pence (d.), and a penny comprised 4 farthings.

The counties and their boundaries referred to in the book are as they existed before 1974.

1

The agricultural revolution

For many years the history of English agriculture since the sixteenth century has been dominated by rival notions of an 'agricultural revolution': a period when changes in agricultural output and sometimes also in the organisation of production are held to be of particular significance. While there is a general consensus that an 'agricultural revolution' involves technological change of some kind, there is no consensus as to what are the significant changes, nor is there any agreement over the chronology of such 'revolutionary' events. Despite this uncertainty, the substantive issues of the debate remain of central importance to understanding the development of both English agriculture and the English economy in the three and a half centuries from 1500 to 1850. This chapter will briefly review the debate before making the case for a set of criteria appropriate for identifying an 'agricultural revolution'.

Agricultural revolutions

Phases of 'revolution' have been identified for at least five periods between 1560 and 1880, and each has been characterised by a different combination of 'significant' agricultural developments. Yet despite these differences there is a remarkable consensus, which stretches back to the earliest writing on the subject, that the essence of the 'agricultural revolution' was an increase in cereal yields per acre, that is the amount of grain that could be produced from a given area of land sown with a particular crop. The mechanism for raising yields was described by William Marshall in 1795: 'No dung – no turnips – no bullocks – no barley – no clover – nor ... wheat.' He was describing a crop rotation that became known as the Norfolk four-course, which was regarded by contemporary agricultural writers as responsible for unprecedented improvements in crop yields and farm output. It is not surprising, therefore, that the set of principles underlying the rotation are regarded by historians as the cornerstone of the 'agricultural revolution'.

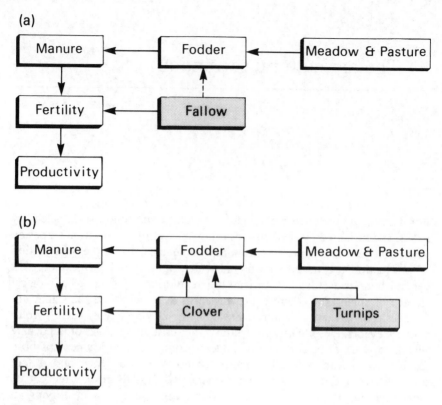

Figure 1.1   The conventional agricultural revolution. *Source:* Overton (1991), 286.

This rotation and its impact will be considered in more detail in Chapter Three, but for the moment we will follow the literature and show how it has been characterised by historians. This is shown diagrammatically in Figure 1.1. Turnips and clover are considered important because they enabled English agriculture to break out of a 'closed circuit' which prevented increases in output other than by extending the cultivated area. The upper part of the diagram, (a) shows the situation before the introduction of the 'revolutionary crops', when the land is cultivated under a crop rotation where two or three grain crops are followed by a fallow. Fallow is land left uncropped, either as a mass of self-seeded weeds, or as bare earth which would have to be ploughed to prevent weeds growing. The fallow was necessary to help the fertility of the soil recover after nutrients had been removed by grain crops. Fertility is also seen as a function of the amount of manure, which in turn depends on the amount of fodder available for animals to eat. Most of this fodder came from meadow and pasture, but a weedy fallow would also have provided some food. The productivity of this system is equated with grain yields per acre, which are determined by soil fertility. Assuming the cultivated area to be fixed, any increase in the proportion of land under grain would reduce the proportion of fallow or pasture, so reducing fodder, fertility and yields. Thus, the increase in grain output accruing from an increase in the area under grain crops would be offset by a fall in yields per acre as a result of the reduction in fertility.

This vicious circle was broken by replacing fallows with fodder crops as shown in Figure 1.1(b). The new crops, turnips and clover, had a fourfold impact. First, they replaced the unproductive fallow with a growing crop. Second, the loss of fodder from the fallow was more than compensated for by the new supplies of fodder which increased manure production and therefore fertility. Third, clover has the valuable property of fixing atmospheric nitrogen into the soil. Nitrogen is a major plant food but cereal crops cannot make use of atmospheric nitrogen and must absorb nitrogen through their roots in the form of nitrogen salts, such as ammonium nitrate. As with other leguminous plants (such as peas and beans) clover converts atmospheric nitrogen into nitrates in the soil; in effect, it makes fertiliser. Thus, the loss of the 'rest' given by the fallow was compensated by the fixing of atmospheric nitrogen by clover. Fourth, turnips acted as a 'cleaning crop'. The fallow period had presented an opportunity to eradicate perennial weeds through ploughing, but turnips achieved the same effect by smothering weeds, and could be easily hoed when sown in rows. Turnips also provided winter fodder for animals since they could remain in the ground during the first half of the winter. Thus, the way was open for an ascending spiral of progress in so far as more food could be produced from the same area of land. There is much truth in this conventional view, but it is an oversimplification and a very partial view of the agronomy of the 'agricultural revolution'. It will be explored in more detail in Chapter Three.

Early accounts of an 'agricultural revolution' stressing these technological changes include Toynbee's posthumous *Lectures on the industrial revolution* published in 1884, and an article in the *Quarterly Review* for 1885 written by R.E. Prothero (who later became Lord Ernle) which was subsequently expanded into a book. Like early writing on the industrial revolution, these accounts captured the popular imagination with their emphasis on particular innovations (which in addition to turnips and clover included selective livestock breeding, and mechanical gadgets such as the seed drill), and the 'Great Men' allegedly responsible for their introduction; as Curtler puts it, 'a band of men whose names are, or ought to be, household words with English farmers: Jethro Tull, Lord Townshend, Arthur Young, Bakewell, Coke of Holkham and the Collings'. These men are seen to have triumphed over a conservative mass of country bumpkins and single-handedly transformed English agriculture within a few years from a peasant subsistence economy into a thriving capitalist agricultural system capable of feeding the teeming millions in the new industrial cities.

These technological innovations were placed firmly in the period 1760–1840, and were seen to be facilitated by changes in the institutional structure of farming, especially parliamentary enclosure, which swept away the open commonfields. The process of enclosure was regarded as important because it removed common property rights, which had hitherto inhibited innovation. It was considered a prerequisite for selective animal breeding in that it prevented the promiscuous mingling of livestock on the commons, and it also allowed the development of the large capitalist farms required for innovations in farming techniques.

Almost all the features of this early depiction of an 'agricultural revolution' are in some dispute, but there is general agreement about two particular criticisms. The first is over the role of the 'Great Men' as pioneers and innovators. It has been shown that 'Turnip' Townshend was a boy when turnips were first grown on his estate, Jethro Tull was something of a crank and not the first person to invent a seed drill, and that although Coke of Holkham was a great publicist (especially of his own achievements), some of the farming practices he encouraged (such as the employment of the Norfolk four-course rotation in unsuitable conditions) may have been positively harmful. Despite this evidence, the myths associated with these individuals have proved extremely difficult to dislodge from literature not directed at a specialist historical audience, including popular histories and texts for use in schools. The second criticism is over the pace of change. Most historians now agree that many of the agricultural improvements ascribed to the 'Great Men' of the eighteenth century, including the innovation of new crops and structural changes like enclosure, had long antecedents and may be traced back into the seventeenth century if not earlier.

Nevertheless, a revisionist view of the traditional story, which owes much to the work of G.E. Mingay, still firmly locates the *revolutionary* period in the century after 1750. It is acknowledged that the picture of a sudden and rapid transformation in the eighteenth century is mistaken, and that eighteenth-century improvements had long antecedents, yet, as the title of Chambers' and Mingay's *Agricultural revolution, 1750–1880* suggests, their revolution is firmly placed in the period advocated by Toynbee and Ernle. Their criteria are similar too. Heading the list are new fodder crops and crop rotations, convertible husbandry, and parliamentary enclosure, with a group of less important elements following: animal breeding, field drainage, and new machinery and implements. Although their book reaches forward into the 1870s, the revolutionary period is seen as the eighteenth century. Chambers and Mingay justify these post-1750 changes as 'revolutionary' because they estimate that an additional 6.5 million people were being fed by English agriculture in 1850 compared with 1750. Although more land was cultivated, much of this extra food was the result of increases in output per acre.

There are two claims for nineteenth-century 'agricultural revolutions'. The first of these takes up a theme first introduced by Darby and argues that there was a revolution on the claylands brought about by tile under-draining which allowed the conventional 'agricultural revolution' to spread from light soils to heavier ones. The second, described by F.M.L. Thompson as the 'second agricultural revolution', is based on the import of feedstuffs such as oil-cake, and fertilisers such as guano, and is seen (despite its title) as the third step in the transition to a modern agriculture. Like the first agricultural revolution its significance lies in breaking a 'closed circuit', only this time by importing inputs to the agricultural system from abroad. On the other hand, the pattern of farming, the mix of crops and stock, and the nature of farm output remained similar in their essentials to the pattern of the late eighteenth century.

The remaining contributions to the debate have all emphasised the importance of earlier periods. The most vigorous onslaught against the revolution of Toynbee and Ernle has come from Eric Kerridge whose *Agricultural revolution* announces, 'This book argues that the agricultural revolution took place in England in the sixteenth and seventeenth centuries and not in the eighteenth and nineteenth.' More specifically he argues that the 'agricultural revolution' occurred between 1560 and 1767, with most achieved before 1673. Kerridge attempts to establish his claim by dismissing the significance of agricultural change after 1750, and by stressing the importance of technological innovation in the earlier period. He has three lines of attack on the significant features of the traditional post-1750 agricultural revolution. First, he argues that some of them did not occur at all – the mechanisation of farming in the eighteenth century for example. Second, he considers some features to be 'irrelevant' (including parliamentary enclosure, the replacement of bare fallows, the Norfolk four-course rotation and selective breeding). Third, he maintains that some technological innovations occurred much earlier, such as the introduction of fodder crops, new crop rotations and field drainage. While some of these points are accepted (farming was not mechanised in the eighteenth century for example), few historians accept his cavalier dismissal of so many features as 'irrelevant'.

After this demolition Kerridge then constructs an argument for an earlier 'agricultural revolution' on the basis of seven criteria which form chapter headings for his book: up and down husbandry, fen drainage, fertilisers, floating the watermeadows, new crops, new systems and new stock. Most emphasis is placed on up and down husbandry (called by others 'convertible' or 'ley' husbandry), which increased fertility by breaking the distinction between permanent grass and permanent tillage and rotating grass round the farm. Once again the criteria are technological, and Kerridge justifies his claim for an 'agricultural revolution' by pointing to the fact that domestic agricultural production coped with a doubling of the English population between 1550 and 1750.

Also writing in the 1960s, Jones considered that from the mid-seventeenth century 'English agriculture underwent a transformation in its techniques out of all proportion to the rather limited widening of the market', and 'If there was a revolutionary phase it ... had come during the Commonwealth and Restoration periods', and his conclusions were reinforced by the work of John. They argue that the importance of the century after 1650 lies in a series of cropping innovations, rather than the feeding of more people, since, in contrast to the following century, population growth remained roughly static after 1650. Innovation led to increases in grain output per acre, and a rise in total output evidenced by rising grain exports. The processes by which output increased are virtually the same as for the post-1750 revolution: a rise in the fertility of the soil through turnips and clover and their associated crop rotations. The stimulus for change is seen as a run of sluggish grain prices which squeezed farmers' profits. This caused them to keep more live-stock, and more important, to lower unit costs of production by raising yields through the innovation of fodder crops. Landlords supported their tenant farmers and encouraged them to make improvements during this period. Jones also argues for a change in the regional geography of farming, since these new methods were most readily adopted on lightlands (princi-pally the chalk downlands of southern England).

Thus by the 1970s, three periods – 1560–1673, 1650–1750 and 1750–1850 – were rival contenders for the period of the 'agricultural revolution' in England. By the end of that decade, the second one was the most popular. The subsequent publication of two volumes of the authoritative Cambridge *Agrarian History of England and Wales* has done nothing to clarify matters. Volume V concludes that for the century after 1650, a depression in grain prices prompted innovation and enterprise, but the full harvest of this inge-nuity in the form of an 'agricultural revolution' was not reaped until after 1750. On the other hand, the succeeding Volume VI, dealing with the period 1750–1850, considers that although the agricultural changes in the century after 1750 were remarkable, 'It could hardly be said that they amounted to an agricultural revolution', since they were a limited prepara-tion for the greater changes yet to come.

In the 1990s, two important contributions have reinforced the view that the 'agricultural revolution' was a phenomenon of the period before 1750. Allen argues that what he calls the 'yeoman's agricultural revolution' occurred mainly in the seventeenth century: 'most of the productivity growth in early modern England was accomplished by small farmers in the open fields during the seventeenth century' and was marked by a 'doubling of corn yields', whereas the eighteenth century saw a landlords' revolution through enclosure which did not increase output but redistributed income from farmers and labourers to landlords. Finally, Clark, in a general discussion of the 'agricultural revolution', has concluded, 'There was no agricultural revolution between the early eighteenth and mid-nineteenth centuries', and offers the extraordinary opinion that 'the finding of little productivity growth in agriculture from 1700 to 1850 is consistent with all of the reliable information we have for agriculture in this period'.

Criteria for an agricultural revolution

Resolution of the question as to whether or not an 'agricultural revolution' took place during a particular period is both a conceptual and an empirical issue. Whether or not agricultural developments in particular periods are interpreted as revolutionary depends on how the concept of an 'agricultural revolution' is defined as well as the evidence from the historical record. Historical evidence by itself cannot reveal the presence or absence of an 'agricultural revolution'. Three sets of criteria can be identified in the literature as implicitly or explicitly constituting the grounds for claiming an 'agricultural revolution'. The first of these embraces a wide variety of changes in farming techniques. These include the introduction of new fodder crops and new crop rotations, the watering of meadows, the improvement of livestock breeds and the introduction of machinery. The second is the fact that English agriculture was successful in responding to the challenge of feeding a growing population, an argument that has been employed for both the sixteenth and early seventeenth centuries, and for the century after 1750. The third is the view that an 'agricultural revolution' is best characterised as an increase in output brought about by improvements in productivity, where productivity is defined as output per unit of input. Indices of productivity vary depending on the combinations of inputs and outputs employed, but the two most important agricultural inputs, and therefore the two most important productivities, are land and labour. Grigg first stressed the importance of productivity change as an indicator of an 'agricultural revolution', and it has been accepted by most recent contributors to the debate, including both Allen and Clark. Earlier writers, however, were rather reluctant to engage with concepts of productivity *explicitly*; their 'agricultural revolutions' are *implicitly* based on productivity change, although their concepts of agricultural productivity are woolly and ill-defined: for example, the productivity of land is often misleadingly equated with grain yields per acre (often for wheat alone) while discussions of the productivity of labour have been subsumed in the issue of the 'release' of labour from the agricultural to the industrial sector of the economy during the industrial revolution.

It could be argued that these three conceptions of an 'agricultural revolution' are rather narrow: they are concerned primarily with changes in the methods and techniques for producing food, with what Marx called the 'forces of production'. A wider notion of an 'agricultural revolution' (an agrarian revolution perhaps) would link these to changes in what he called the 'relations of production', which other writers sometimes refer to as institutional change. These issues are concerned with the establishment of private property rights in land, the replacement of feudal tenures and estates with leaseholds for a period of years, changes in the size of farms, and changes in the ways in which people were employed by others on the land.

The argument adopted in this book is that the agricultural sector of the economy underwent two related transformations during the period 1500–1850. The first, which was largely a product of the period after 1700, enabled English agriculture to increase output to unprecedented levels, and sustain that output in the face of ever-increasing demand. Before the eighteenth century, English population seemed unable to exceed a maximum of about five to six million people. When population reached this maximum (in the early fourteenth century and again in the mid-seventeenth century) it stopped growing. A strong case can be made that this limit was set by the inability of agricultural output to expand. Whenever population grew, food prices rose too, indicating that supply was not keeping up with demand. Yet from the mid-eighteenth century population breached this maximum and continued to grow rapidly thereafter; moreover, the link between population growth and the growth in food prices was irrevocably broken. While output was being raised to unprecedented levels, the amount of food produced by each worker in agriculture also rose. In 1500 roughly 80 per cent of the population were working in agriculture; by 1850 the proportion was around 20 per cent. This rise in the productivity of labour in agriculture meant that an increasing proportion of the population were working in the industrial and tertiary sectors of the economy; in other words that an industrial revolution was taking place.

These rises in agricultural output and in labour productivity are of crucial importance, not just to the history of agriculture but to any broader history of England. So too are the social and economic relationships within farming which were the subject of the second transformation. In the early sixteenth century, around 80 per cent of farmers were only growing enough food for the needs of their family household. By 1850, the majority of farmers produced much more than they needed for themselves, and were businessmen farming for the market. Markets and marketing had been revolutionised; private property rights were universal, and farming was dominated by the tripartite class structure of landlord, tenant farmer and labourer. The period during which these changes occurred was a more protracted one, and unlike the first transformation there are strong grounds for claiming it was underway by the mid-seventeenth century. The significance of these changes, which amount to the establishment of agrarian capitalism, lies both in their effects on production, and in their impact on the lives of those working in the countryside.

The distinction between changes in agricultural production, and in the social and institutional framework in which that production was carried out, echoes a division in the study of English agrarian history that extends beyond the debate over an 'agricultural revolution'. So the framework adopted here enables the incorporation of other important themes in the historiography of English agriculture. On the one hand there are studies that concentrate on farming practice, stemming from the 'Leicester School' of agricultural history. This work (often labelled 'cows and ploughs' agricultural history although very little work has been specifically carried out on either), is concerned with the practicalities of farming and especially with regional differences in agricultural systems. It developed after the Second World War in conjunction with the exploitation of the many new sources for agrarian history that were becoming increasingly available in newly established county record offices, and its influence is clearly evident in Volumes IV and V of the Cambridge *Agrarian History*.

On the other hand there are studies of social and institutional change, which have a longer pedigree. This tradition dates back to Marx, who saw the parliamentary enclosure movement of the eighteenth and nineteenth centuries as the mechanism responsible for creating the English proletariat. In the first decade of the twentieth century several studies tackled the origins and consequences of capitalist agriculture, including Tawney's *Agrarian problem in the sixteenth century*, and the Hammonds' *Village labourer*. A renewed interest in social and institutional matters occurred in the 1970s, associated with the revival of social history and a resurgence of interest in Marxism, and is exemplified by Robert Brenner's article on agrarian class structures in *Past and Present*, and by a new journal, *Rural History*, founded in 1990 explicitly to cater for the social and cultural aspects of rural history.

Thus, while this book is concerned with the issue of the agricultural revolution, it also deals with some of the broader issues in English agrarian history. This is especially the case with the next chapter, on farming in the sixteenth century, which is not about transformation at all, but introduces English farming and rural society in the period before any transformations took place. It also addresses an important theme of the 'Leicester School': the regional variety in farming, and rural economies more generally. Chapter Three is devoted to the transformations in output and productivity, while Chapter Four attempts to unravel transformations in institutional and structural change. Finally, Chapter Five returns to the themes of this chapter to reconsider the 'agricultural revolution'.

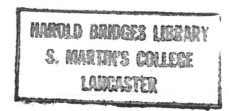

2

Farming in the sixteenth century

An understanding of the rural world of early modern England must start
with the activity that occupied most of the population: the practical business of farming. This chapter begins with a description of farming operations and considers the constraints farmers faced in attempting to maintain
or increase their output of crops and livestock. Farming was not a uniform
activity, so some of the differences between farming enterprises are
discussed in terms of their products, labour requirements, income flows
and relations with the market. The chapter then investigates the land being
farmed in terms of patterns of ownership and rights to property. The next
section moves from farms to farmers, looking at their social status, and at
their social and economic relationships within the local community.
Finally, the chapter explores the relationships between these various
elements of the rural economy within the context of a framework of rural
regions.

Farming

Five hundred years ago English farmers grew four major cereal crops:
wheat, rye, barley and oats, together with the pulse crops of peas and
beans; they also kept cattle, sheep, pigs and poultry. Although these crops
and livestock are kept by farmers today, cereal crops have been changed
dramatically by plant breeding during the twentieth century, and livestock
characteristics have been transformed by selective breeding. During the last
five centuries sugar beet, potatoes, rape, turnips and swedes amongst
others have been added to the sixteenth-century list of crops, although all
but sugar beet were introduced in the three centuries before 1850. In the
early modern period arable land was prepared by ploughing and
harrowing, fertiliser of some sort was added, the seed sown, the crop
protected during its growth (by weeding and pest control) and finally
harvested and prepared for consumption. These basic operations of arable

husbandry are carried out by farmers today, but have been revolutionised by new implements and machines, new sources of power and the use of chemicals. The introduction of machinery began in the early nineteenth century and it was also during this period that the range of fertilisers available to farmers started to expand as a consequence of imports from abroad; until then the main source of fertiliser was animal manure. In the twentieth century imports of natural products have been replaced by 'artificial' fertilisers manufactured using fossil fuels. Since the Second World War crop protection has been revolutionised by the development of herbicides and pesticides which control weeds, plant diseases and pests, with relative ease. Until the twentieth century farmers supplemented their own labour with that from animals (horses and oxen) as the source of energy for almost all their farming tasks. Animals pulled the plough, the harrow and the roller; they hauled the cart and the waggon; and they walked round in circles in the horse gin to provide the power for grinding corn. Steam power began to supplement animal power in the nineteenth century for some tasks but it is only in the second half of the twentieth century that the internal combustion engine has replaced animals as the main source of power on the farm.

The fundamental differences between arable and livestock farming make it convenient to discuss each separately, but it should be evident that almost all farms had to combine crops and livestock in the early sixteenth century, if only because animals were needed for draught and as a source of manure. Whereas today a farm can be exclusively arable with no livestock at all, before the twentieth century arable farming at its most arable was mixed farming. Although the balance between the various crops and livestock types varied considerably from farm to farm, in comparison with farms of today, the vast majority of farms in the sixteenth century were mixed farms.

Arable and pasture: crops and farming operations

The arable farming year began after the harvest in August or September when the first task was to plough the land. Most ploughs had a mouldboard which turned over the soil, burying stubble and weeds, thus leaving bare soil on the surface. Land broken up from pasture was ploughed up to five or six times, but the land that had been under an arable crop or a fallow was usually ploughed up to four times, depending on the state of the soil. The land was often ploughed in a series of ridges and furrows to help drainage, sometimes with a height of several feet between the top of a ridge and the bottom of a furrow. After ploughing, harrowing the soil would help make a tilth on which the seed could be sown, but if the crop was not to be sown until the spring, the plough clods would be left to be broken down by winter frosts before being harrowed in the spring. Oxen (castrated adult male cattle) were traditionally used to draw the plough, but from the middle ages onwards they were being replaced in some parts of the country by horses. Depending on the nature of the land, the plough would be pulled by a team of up to eight oxen or from one to four horses. The so-called winter cereals, wheat and rye, would then be sown in September or October, as soon as the soil was ready for them. Wheat and rye seed were sometimes mixed before they were sown to produce a crop which went under a number of names including maslin, or mixtlyn. The so-called spring crops, barley (or the six-rowed bigg in the north of England), oats, peas and beans, would be sown from late February to April when the soil was warming up and an adequate tilth could be obtained. Seed was sown by 'broadcasting', a process by which the sower walked across a field scattering the seed by hand from a basket hung round his neck. Finally the seed would be covered by harrowing again. Once the crop was growing, the main task was to keep it free from weeds, or at least minimise the competition from weeds. This was a labour-intensive job and the extent to which it was undertaken would depend on the availability and cost of labour.

Today, cereals in England are harvested using a combine harvester. This cuts the corn, threshes it, stores the grain in a tank, and discards the straw. The straw is then usually bound into bales by a baler before being cleared from the field. The process was more protracted in the sixteenth century. Cereals were usually harvested during August and September, with the winter grains being harvested before the spring-sown crops. Wheat and rye were 'reaped' with a small hook called a sickle while barley and oats were 'mown' with a much larger scythe. A sickle was held in one hand as the reaper (who could be a man or woman) bent over the crop and used the other hand to hold the corn while sawing through the straw. A mower (almost always a man) stood upright and slowly walked through the crop slicing the scythe with a smooth rhythm. Once the corn had been cut, it was bound into sheaves by workers following the reaper or mower. These sheaves were then stacked into stooks which could stand in the field for a week or more to dry. The stooks would then be carted to be stored in a barn, or built into a stack which would be thatched with straw to keep out the rain. The final field operation of the harvest year was gleaning, the process whereby the field was scoured for grain that had been spilt during harvesting. Individual grains were laboriously picked up, usually by women and children who, in the sixteenth century at least, were able to keep their gleanings as a perquisite. The tools and implements involved in this farming activity, reproduced from a late seventeenth-century farming treatise, are listed in Table 2.1.

Grain was threshed (separated from the ear) as it was needed

Table 2.1 *Implements of husbandry for a seventeenth-century farm*

'He that goes a borrowing, goes a sorrowing' hence the necessary items for a husbandman, in addition to ploughs, spades, hoes and carts, are:

Belonging to the Arable and Field-land

Harrows
Drags
Forks
Sickles
Reap-hooks
Weed-hooks
Pitch-forks
Rakes
Plough-staff and Bettle
Sleds
Roller
Mold-Spears and Traps
Cradle-sythes
Seed-lip

To the Barn and Stable

Flails
Ladders
Winnowing-Fan
Measures for Corn
Sieves and Rudders
Brooms
Sacks
Skeps or Scuttles
Bins
Pails
Curry-Combs
Main-Combs
Whips
Goads
Harneys for Horses, and
Yokes for Oxen
Pannels
Wanteys
Pack-saddles

Suffingles
Cart-lines
Skrein for Corn

To Meadows and Pastures

Sythes
Rakes
Pitchforks and Prongs
Fetters and Clogs, and Shackles
Cutting spade for Hay-reeks
Horse-locks

Other necessary Instruments

Hand-barrows
Wheel-barrows
Dibbles
Hammer and Nails
Pincers
Sissers
Bridle and Saddle
Nail-piercers or Gimlets
Hedging-hooks and Bills
Garden-sheers
A Grindstone
Whetstone
Hatchets and Axes
Sawes
Beetle and Wedges
Leavers
Shears for Sheep
Trowels for House and Garden
Hod and Tray
Hog-yokes and Rings
Marks for Beasts and Utensils
Scales and Weights
An Aul, and every other thing necessary.

throughout the year. Sheaves of corn were spread on the barn floor and beaten with a wooden instrument called a flail, which consisted of two pieces of wood linked together by a universal joint. Although it sounds rather crude, threshing was a skilled job because the grain had to be extracted from the ear without damaging the straw. Grain was winnowed (separated from the chaff) by shaking it in a wicker basket (or fan) so that the wind would blow the chaff away, and weed seeds were then separated from cereal grains by means of a sieve or corn screen. The grain could then be put into sacks ready to be ground into flour on the farm, taken to a local miller, or carted to market to be sold.

Bread made from wheat flour was the most highly regarded, but was only eaten regularly by about half the population in the early sixteenth century. The main wheat growing areas were in the south of the country, but people in the north could enjoy a diet of wheaten bread provided they could afford it. Barley could also be made into bread, but most of the barley crop was turned into malt which was then used to brew ale or beer. It was also possible to feed barley (but not wheat) directly to animals provided it had been softened by soaking in water or by rolling. Oats formed the main component of diets in the north and west of the country, where the crop dominated, but in other parts of the country its main use was as fodder for horses. Other fodder crops, such as hay, peas and beans, were usually retained for use on the farm, although they were sometimes marketed. Overall perhaps as much as 20 per cent of arable production was consumed by livestock. No national figures of the proportions of the major cereals grown are available until the nineteenth century (Tables 3.13–3.15) but data from probate inventories (Table 3.12) give some idea of crop proportions in various parts of the country.

During the growing season animals ate grass, and in the winter they ate hay cut from meadows in the summer. The hay harvest usually took place in June and July. The grass was mown with a scythe and left to dry; it was then raked into cocks, and carried to the barn or made into a rick, which would be thatched to prevent rain rotting the hay. Little else was needed for the maintenance of grassland except perhaps for removing the worst weeds like thistles and docks and dealing with pests, particularly moles. Livestock needed much more attention. The annual cycle began in the spring when most animals were born, so that their mothers could take advantage of new grass to produce milk, although farmers arranged for calves to be born at other times of the year to ensure continuity of milk production. Calves and lambs then had to be weaned, and those males not required for breeding, castrated. In September and October bulls would service cows and heifers and rams would service ewes at roughly the same time. Sheep also had to be washed and clipped in June. Day to day activities involved milking both cows and sheep, and the subsequent processing of the milk into cheese and butter, which was women's work in the dairy; and in the winter providing cattle with fodder. Even when feeding outside in the summer both sheep and cattle had to be watched to ensure they remained where they were supposed to be. Sometimes cattle were tethered, and sheep were often 'folded' on arable land from May to October. The fold was a temporary enclosure made of wooden hurdles which concentrated the flock on specific areas of the arable, usually the fallow. The length of time animals were kept on the farm varied considerably. Ewes in a breeding flock could remain for six or seven years, and dairy cattle for even longer. Cattle kept for fattening might remain on the farm for up to four years before slaughter, although some farmers specialised in buying in young cattle, fattening them over the summer, and selling them in the autumn. Likewise some farmers might purchase lambs in the spring and fatten them up for sale in the autumn of the same year.

While the majority of farmers were mainly concerned with the cultivation of cereals and the husbandry of sheep and cattle, they would also be engaged in the production of other commodities. Depending on local circumstances, farmers might have an area of woodland, an orchard and a kitchen garden; they would keep bees for honey, and perhaps pigeons in addition to the ubiquitous pigs, geese, hens and ducks. These involved additional work and there were more tasks which had to be fitted in as and when time was available. Hedges and ditches had to be maintained, buildings repaired, wood cut for fuel, and in certain areas arable fields improved by removing large stones, a back-breaking job often done by women and children.

Many of the crops and livestock mentioned so far provided raw materials for industry as well as food. Agriculture provided virtually all the raw materials for clothing and other textiles, transport and furniture, and made important contributions for lighting, writing materials and building. Clothes were made from wool, linen and hides; tallow (fat) from sheep and cattle was used for lighting; people wrote with a quill pen made from a goose feather on writing materials made from calf skins (vellum) and sheep skins (parchment). All forms of transport (sledges, carts, waggons, boats and ships) were made from wood, and land transport was powered by animals fuelled by crops. A few farmers grew more specialist industrial crops on a commercial scale. These included hemp to make ropes, flax to be turned into linen, and, from the mid-sixteenth century, hops to flavour beer. More esoteric crops included woad, weld, madder and saffron to make dyes; and teasels to prepare wool for spinning.

Constraints on agricultural production

Late twentieth-century farmers are able to overcome three problems (albeit at some cost), that were almost insuperable for early modern farmers: increasing soil fertility, controlling pests, and controlling plant and animal diseases. All sixteenth-century farmers faced the problem of maintaining the fertility of their soil: in general terms replacing the nutrients removed by growing crops. More specifically, the three main elements needed for crop growth are nitrogen, phosphorus and potassium, and the most important of these in the early modern period was nitrogen. This was the 'limiting factor' in that it was the element most likely to be in short supply and therefore limit the growth of the crop. By the mid-nineteenth century much of the old arable land in Britain appears to have lost two-thirds of the soil nitrogen that was present before farming began, and so the key to maintaining, let alone increasing, crop yields was the management of soil nitrogen. Nitrogen can only be utilised by plants in the form of mineral nitrate salts, and in the early modern period the main source of these was the decay of organic nitrogen from plant materials. Some of this recycled nitrogen came from plant residues ploughed into the ground: from stubble, weeds and grass. Organic nitrogen is recycled more rapidly through animals and so one of the most important sources of mineral nitrogen was animal manure, but it is important to realise that animals do not make manure out of nothing, they are merely processing the nutrients contained in the plants they eat.

Maintaining soil fertility involved two processes: first, conserving existing supplies of nitrogen, and second, facilitating the addition of new supplies of nitrogen into the farming system. Early modern farmers were, of course, ignorant of the existence of nitrogen, but they were nevertheless aware of strategies to maintain fertility which, although they did not realise it, involved the conservation of nitrogen. Thus farmers were very aware of the value of manure, and went to some lengths to conserve their stocks and deposit them where they were most needed. A particularly important process was grazing animals by day on permanent pasture and then folding them on the arable (usually the fallow) at night. Since sheep and cattle eat mostly during daylight but urinate and defecate equally during the night and day, moving the animals in this way results in a movement of crop nutrients from pasture to arable. Folding sheep was essential to the maintenance of arable husbandry in many light-soil areas of the country. Aside from manure and crop residues farmers were also aware that adding lime or marl to the soil would improve fertility. It did this not because these materials provide crop nutrients directly, but because they reduce soil acidity and improve soil structure enabling the bacteria breaking down manure and crop residues to work more effectively and make more nutrients available to growing crops.

In the sixteenth century the bare fallow provided one means by which new nitrogen was added to the soil. During a period of fallow, soil bacteria will convert atmospheric nitrogen into soil nitrogen leaving new reserves for the next crop, provided rainfall does not leach the nitrate salts away. Usually a more effective mechanism of transferring atmospheric nitrogen into soil nitrogen was through the cultivation of legumes, a class of plant that facilitates the conversion of atmospheric nitrogen into soil nitrogen through bacteria attached to plant roots. In the sixteenth century the main legumes cultivated were peas and beans, but, as we shall see in the next chapter, the widespread cultivation of another legume, clover, was to revolutionise farming by replenishing nitrogen at an unprecedented rate.

The fallow was also necessary to help in the control of weeds, pests and pathogens. Modern farmers can deal with these using herbicides, pesticides and fungicides, but in the early modern period farmers were comparatively helpless in combating these problems. Repeated ploughing during the fallow period was the only way to eradicate persistent perennial weeds. Annual weeds were less of a problem and could be removed from the crop as it was growing, although this could be expensive in terms of labour and could damage the crop. The range of pests confronting farmers has not altered much in the last five hundred years and in some cases the ways of dealing with them have not altered either. Birds are still scared off growing crops by a loud noise for example, even though the means of making the noise have changed. It is estimated that a fifth of the world's grain output is currently lost to pests and diseases and it is obvious that such losses must also have been considerable in the past, perhaps reaching a third in early modern England. Crop rotation helped prevent the build-up of both pests and disease-causing pathogens, as did the careful selection of disease-free seed. Rotations varied widely across the country and from farm to farm, depending on the nature of the soil and climate, but also on local custom and tradition. The most important crop (often wheat) was sown after the fallow or a pulse crop of peas or beans. Several cereal crops would then be taken in succession, often with barley following wheat, and oats following barley, before the land reverted to fallow again. Aside from rotation a variety of methods were adopted to combat diseases, the most prevalent of which was the steeping of seed (in brine, lime, blood or urine for example) prior to sowing, and reducing soil acidity by the addition of lime. Stored grain was particularly susceptible to attacks from vermin. Staddle barns were one of the most effective ways of resisting attack: granaries were raised on mushroom-shaped stone legs which were impossible for rats and mice to climb since they were defeated by the overhang. Animals also succumbed to disease and although a great body of lore existed for the treatment of animals, in general it was no more effective than was medical treatment for human beings.

Modern farmers are just as dependent on the vagaries of the weather as were their sixteenth-century counterparts, but different technologies of farming meant that the latter were affected in different ways. They faced more problems at seed time than modern farmers because their land was relatively poorly drained. Today, most of the farmland of England is underdrained and therefore the appearance of the current landscape is a misleading indication of the state of drainage in the past. Successful underdraining on a large scale had to wait until the nineteenth century with the introduction of the tile drain. Before then, ridge and furrow was the principal means of surface drainage, but from the seventeenth century onwards hollow drains seem to have been more frequently employed, whereby stones or bushes were put into trenches and covered with soil. Nevertheless, little effective underdraining had been carried out by 1800. However, although the land was wetter in the past it did not suffer damage from heavy machinery: land that today is too wet to take a tractor and seed drill could, in the sixteenth century, take a man broadcasting seed. High moisture levels at harvest time were less of a problem than they are today as grain could dry as it stood in the sheaf before being brought home from the field. In general, livestock farmers were more susceptible to bad weather than their modern counterparts. The absence of a treatment for foot-rot in sheep, for example, meant that farmers keeping sheep feared wet weather, while those with cattle feared drought in the absence of alternative fodder to grass.

Varieties of farm

It should be evident from the variety of agricultural products and farming techniques already described that there were many different types of farm. Regional variations in farming types will be considered later in this chapter, but some of the major characteristics of the various farm enterprises will be considered here. The description of farming tasks at the start of the chapter showed that the pattern and intensity of farm work differed for arable and pastoral production. The daily rhythm of work on an arable farm, be it ploughing, weeding, harvesting or threshing, was fairly constant throughout the working day. On the other hand, the work involved in tending livestock was concentrated at certain times of the day. Dairy farmers, for example, were tied to a morning and an evening milking, but had little or no field work outside the hay harvest, and thus had more time on their hands during the day than did arable farmers. This gave them the opportunity to engage in other activities such as the production of handicraft commodities. The demands on the time of the arable farmer were also more constant throughout the year. There was a peak of work in late summer with the harvest, but there was always plenty of work during other seasons of the year. Pasture farmers had two seasons of peak labour demand, with lambing and calving in spring, followed by the hay harvest in early summer; but for most of the rest of the year they were free to engage in other activities if they so wished. Income flows also varied between pasture and arable farms. Pasture farmers had their working capital immediately realisable, whereas arable farmers, or at least those for whom cereals were the main cash crop, had to wait until the harvest before they could sell their crop and realise some income.

The ability to realise income also varied with farm size. Farmers with a small farm, with few reserves, could not afford to hold stocks of grain and were forced to sell their crop soon after the harvest. This meant they usually received the lowest prices, for while the demand for grain was constant over the year, it was in greater supply immediately after the harvest so its price was usually at its lowest. Larger farmers, on the other hand could afford to wait until prices rose, in effect speculating on the future price of grain. As the Berkshire farmer Robert Loder put it in 1621: 'it is good husbandrie at all times of the yeare to marke well ye likelines of the subsequent yeare for plentie or scarcetie, & therupon to kepe or sell accordinglye as I shall think it likely to fall out most for my proffit'. The income flows of livestock farmers were much more varied and depended on the nature of their livestock enterprise. They could, however, sell animals at any point in the year if they had to, and did not have to wait for the harvest to raise cash if they were in a tight financial corner. Specialist cattle farmers might concentrate their purchases and sales at particular seasons of the year. Peter Temple, a substantial Warwickshire farmer in the sixteenth century, bought lean beasts in the spring, mostly from Wales, fattened them up, and sold them off from midsummer to December. On the other hand, his sheep farming enterprise involved a self-sustaining flock: no sheep were bought in and sales were more evenly distributed over the year. The pattern of income flows was different again for dairy farmers. Once a herd had been established income would flow fairly evenly throughout the year from the sale of dairy products, although sales of male calves, barren cows and female calves not needed for replacements would be concentrated in the spring, since they would be bought by farmers who would fatten them up before slaughter.

The extent to which farmers could vary the quantity or the nature of their output in response to changes in the prices of agricultural products (their elasticity of supply) depended more on the size of their enterprise than on whether it was geared towards crops or livestock. Generally speaking the supply of farm products was inelastic (so a change in price led to a less than proportionate change in output) because neither arable nor livestock farmers could increase their production quickly in response to short-term price changes. Arable farmers had to wait almost a year between sowing winter corn and selling it, while livestock farmers could wait up to four years before beef cattle were ready for slaughter. If cereal farmers were holding stocks of grain on the farm, they could, however, sell them when prices rose, and large livestock farmers could afford to sell some of their existing stocks.

The most important determinant of prices in the short term were changes in the quantity of grain and livestock products supplied. For cereal crops this was almost entirely dependent on the size of the harvest, and for livestock products on the number of animals, which was in turn dependent on the state of fodder supplies and the incidence of animal disease. Thus in the short term when prices were high it was because output had fallen and, conversely, when prices fell it was because output had risen. However, the proportionate change in price was greater than the proportionate change in output reflecting the inelastic demand for grain. Thus in bad harvest years the reduction in the quantity of grain available for farmers to sell was more than offset by the rise in the price of grain, so a farmer's income could actually be above that of a normal year. Conversely, when the harvest was good, the advantage of having more grain to sell was more than offset by the reduction in price. As the porter in Macbeth puts it, 'Here's a farmer that hanged himself on the expectation of plenty' (Act II, Scene I). Small farmers, on the margins of subsistence, had a very different experience. If the harvest was good they would have a surplus to sell although prices would be low. On the other hand, if the harvest was a bad one, and failed to provide enough food for their subsistence needs, they were forced to buy on the market when prices were high. In other words they were forced to buy dear and sell cheap. This is demonstrated in Table 2.2 which compares two hypothetical farms, one with 100 acres of wheat and the other with 10 acres. In a normal year the yield of wheat is 10 bushels per acre and so gross output is 1000 bushels and 100 bushels for the two farms. Seed for the next crop and on-farm consumption are deducted from this leaving a net output of 700 and 25 bushels respectively, which, with wheat at 10 pence per bushel gives a total revenue of £70 and £2.50. The relationship between the proportionate change in output and the proportionate change in price is not constant, but is given by the formula

$$y = \frac{0.757}{(x - 0.13)^2}$$

where x is output and y is price. The table shows that for the farmer with 100 acres of wheat total income *rises* as yields fall. For the smaller farmer income falls as yields fall, and when the harvest is 50 per cent of normal the loss is potentially disastrous.

Table 2.2 *Harvest yield and income by farm size*

Harvest as prop. of normal	Acres	Yield (bushels/ acre)	Gross output (bushels)	Seed (bushels)	Con- sumption (bushels)	Net output	Price (pence/ bushel)	Total income (£)
1.5	100	15	1500	250	50	1200	4.0	48.00
1.2	100	12	1200	250	50	900	6.6	59.40
1.0	100	10	1000	250	50	700	10.0	70.00
0.8	100	8	800	250	50	500	16.9	84.50
0.5	100	5	500	250	50	200	55.3	110.60
1.5	10	15	150	25	50	75	4.0	3.00
1.2	10	12	120	25	50	45	6.6	2.97
1.0	10	10	100	25	50	25	10.0	2.50
0.8	10	8	80	25	50	5	16.9	0.85
0.5	10	5	50	25	50	−25	55.3	−13.85

The attitudes and behaviour of farmers producing exclusively for their own needs were very different from those farmers trying to make a profit. They valued their produce in terms of what use it was to them rather than for its value for exchange in the market. Their decisions on what crops to grow and what animals to keep, whether made individually or collectively, were influenced in part by the nature of the land on which their farm lay and their local climate. However it was often the case that these natural endowments were equally suited to several farm enterprises. Good pasture land for example could support cattle for rearing, dairying or fattening. Some land, including the clay vales of southern England, could equally well be used to grow cereals, or to grow grass to feed livestock. Thus in many situations, if the market was not the major influence on what was produced, the actual mix of crops and stock was determined by local custom and tradition as well as by subsistence needs. Although the market did not have much influence over their production decisions, farmers producing at subsistence levels went to the market to buy the few necessities they needed, to sell surplus corn in a good year, and, in a poor year, to buy corn if their harvest fell below their subsistence needs. Larger, profit orientated, farmers were still constrained by soils and climate, and by local customs and traditions, but also had an eye to the market as to which crop and livestock combinations would make them most money.

It is thus important to try and estimate the balance between farmers producing for exchange and those producing for use or subsistence. It is also important to distinguish between subsistence and self-sufficiency. Subsistence farmers were those producing just enough food for their own needs, although this does not mean that they consumed this food themselves, in other words that they were self-sufficient. In the early sixteenth century no farm could possibly be classified as completely self-sufficient, providing all the material needs of the farm family. Many farmers had to buy things they could not produce themselves; but in any case many would need a cash surplus, if not for rent then for taxes. At the very minimum a family farm employing no labour would have bought salt, and any commodity made from metal, assuming the farm household could construct its buildings, make all its own clothes, furniture, cooking equipment and farm equipment. Subsistence farmers produced no more than was necessary for their subsistence, and were not producing food for its exchange value in the market. However, they would have been trading in the local market and may not have directly consumed what they produced.

Using this definition we can only guess at the proportion of subsistence farmers in the early sixteenth century. Roughly three quarters of the population was engaged in farming around 1520, so on average each agricultural family was producing food for themselves and one third of the requirements of another family. This average is misleading however, since a wealth of local examples based on estate surveys demonstrate that the distribution of farm sizes was skewed. It is likely that quite a high proportion, perhaps around 80 per cent, of farmers were living at subsistence levels at the start of the sixteenth century. The predominance of subsistence farming is also indicated by a relatively high degree of homogeneity in farming from place to place in comparison with subsequent centuries. The density of markets was also higher in the sixteenth century than in the seventeenth and eighteenth centuries, suggesting a less developed national marketing system and by implication a higher degree of subsistence farming. The extent of subsistence farming varied across the country. In the north and west for example, where good arable land was relatively scarce, it was more difficult to produce enough corn for subsistence, and grain was more likely to be brought in from outside. By contrast, in the Midlands livestock and crop husbandry were more integrated, in most areas locked together by a regulated commonfield system, and it is likely that at the start of the sixteenth century most villages would be able to support their population in a normal year.

Field systems

Most farmers in early modern England were subject to the constraints of the field system of which their farm was a part. The term 'field system' refers to the layout (the fields) and the organisation (the system) of the land in a farming community. The layout of fields refers to the disposition of the physical features of the field system, while the organisation of the field system consists of two aspects: the rules and regulations governing how the fields were cultivated, and the legal property rights attached to the ownership and use of land. Understanding field systems is difficult, not only because there were so many varieties of field system – seemingly innumerable combinations of topography, property rights, and farming regulations – but also because most of this variety is now lost. Field systems are also complicated because the relationships between their various elements were not always consistent from system to system. Thus while in some cases the presence of one feature was associated with another (subdivided fields with common property rights for example) in other cases it was not. Medieval and early modern field systems are so enigmatic and alien to our modern western world that their origins are the subject of an unending debate amongst historians and geographers; fortunately, by the turn of the sixteenth century most types were already in place, so that particular debate can be avoided here. Given these complexities, it is not surprising that some of the literature on field systems is also inconsistent, particularly in its use of terminology, so it is important to maintain the distinction between the three elements of field systems: topography, property rights, and the regulation of farming activity.

The topography of the field system refers to the features of the system that were visible in the landscape: the distribution of fields, field boundaries, and subdivisions within fields. In the early sixteenth century many fields looked as they do today; rectangular, surrounded by hedges, ditches or walls, with the whole field under the same crop. As well as these fields, which contemporaries called 'closes' or referred to as 'enclosed', there were much larger fields, perhaps up to several hundred acres in extent, subdivided into long strips of land forming the units of land ownership. Sometimes these strips were separated by a grass strip (called a 'baulk') but often there was no obvious physical boundary between the strips. Fields were often ploughed in a pattern of ridges and furrows to assist drainage, and each strip could correspond to a single ridge and furrow. Strips were often grouped together into units called furlongs or lands, but a subdivided field might also have little closes within its boundaries. Subdivided fields were crisscrossed by networks of paths and tracks giving access to the strips, and many fields had patches of unploughed land under grass or even scrub on wet or boggy parts. These subdivided fields are sometimes referred to as open-fields, which reflects how they looked, but the landscape of sub-divided fields was called 'champion country' by contemporaries.

There are remarkably few pictures of open-fields, but the topography of many field systems can be reconstructed with the evidence of contemporary maps. However, an equally important feature of field systems remains invisible. Today most land is subject to private property rights. This means that exclusive rights to the ownership of a piece of land also give exclusive rights of use. If a farmer owns a field with private property rights he has the exclusive right to use it. No other person has a legal right to use the field, to graze their cattle on it for example, without his permission. In the sixteenth century much of the land of England was not subject to private property rights, but to common property rights. If land was subjected to common rights, exclusive rights of ownership did not give exclusive rights of use. Thus even though an individual owned a parcel of land, other people living in the community could have specific rights to use that land in certain ways. For pastures this could mean they had the right to graze their animals, for woodland the right to gather fuel (the right of estover), and on arable land the right to graze their animals on the stubble after the harvest (known as the common of shack). Some land remains subject to common rights today: it is a popular misconception that 'commons' or 'common land' belong to no-one, to the general public, or to the inhabitants of a village. But common land has nothing to do with common ownership: such land is owned in the same way that other land is owned, by an individual or an institution, but common rights still exist over it. Today these rights are usually rather limited (the right of 'air and exercise' for example) although there are still areas where certain people have the right to graze livestock on common land. Land under common rights was also referred to in the sixteenth century as common land, and, where the land was arable, as commonfields. Land under private property rights was usually referred to as 'several' or as enclosed land.

Thus the term 'enclosed' was used to refer both to the topography of a particular field system and to the existence of private property rights. This implies some connection between the form of the field system and its function, and indeed this was the case. Enclosed fields were usually (but not always) under private property rights and subdivided fields usually (but not always) under common rights. Form and function were also linked in the manner in which farming was organised. Fields grouped together as 'ring fence' farms under private property rights would be managed by individual owners or tenants who could farm as they pleased (although if they were a tenant they might be restricted by the terms of their lease). Both the crops and stock making up the farm enterprise, and the techniques and management of husbandry operations, were under the control of a single farmer, since with no common property rights existing over the land no-one else had an interest in what happened on the farm.

Where subdivided fields under common rights prevailed (a commonfield system) the situation was very different. The typical farm would not consist of a contiguous group of fields or even strips, but would be composed of strips scattered throughout the subdivided fields. Farmhouses would be located in the centre of the village, and have a small area of closes attached to them for the production of vegetables and fruit, and other crops that would not fit in with the constraints of the field system, such as hemp and flax. Animals would be grazed on common pastures, in the common meadow, and on the arable stubble after the harvest. Farmers would also have rights to gather fuel from woodland and other parts of the village. Form and function were also linked in that the large subdivided field was often the unit of rotation so that a three-field system would have a three-course rotation of winter corn, spring corn and fallow.

Many farming operations were carried out by farmers collectively. They would plough together (a process termed co-aration), harvest together, cut hay together, and share in tending one another's animals on the common. Animals would graze on common pastures, and be tethered on patches of grassland in the subdivided fields, but after the harvest, all the animals of the village would be allowed to graze the stubble until it was time for the land to be ploughed, first on the winter corn stubble and then on the spring corn field. A much smaller number of livestock, usually sheep because of the way they eat, would be left on the fallow to clear the weeds. Even those who neither owned nor rented land in the village might have the common right to graze a limited number of livestock simply by virtue of being resident in the village. Winter keep for livestock was provided by hay from meadows, which were often low-lying and too wet to plough for crops. Meadows were usually divided amongst farmers by drawing lots before the hay harvest. Animals were usually excluded from the meadows until after the hay harvest when they might be allowed to graze on the 'aftermath'. In the sixteenth century pigs were likely to be found in areas of woodland or also on the waste, rooting around for food, although a few may have been housed and fed on household scraps and waste from dairies.

Livestock are the key to understanding the nature of the commonfield system because to allow free-range grazing on the stubbles and fallows, and to avoid the need for hurdles or fences, all the strips within a group of furlongs, if not the entire field, had to be sown and harvested together. Economies of scale were also important in folding sheep on the arable since the benefits from a large flock were proportionately much greater than those from a small one, and it has been suggested that commonfields would survive longest where the folding of sheep was an integral part of the field system.

This degree of cooperation obviously necessitated a set of rules and regulations for the running of the field system and effective sanctions for those who broke them. These governed the timing of husbandry operations, such as ploughing, sowing, and the opening of stubble fields to livestock; they controlled the number of livestock that any individual might pasture on the commons (known as a stint); and they stipulated penalties for those who ploughed areas of the open-fields that were supposed to be kept under grass, or who failed to keep ditches clear. When the manor and the field system were territorially the same these rules and regulations were administered by a manorial court, but when they were not, where there was no manor or several manors in the village for example, a village meeting would run the field system. Each village community was autonomous in the way it organised its own field-system: the community made their own decisions in the light of their own circumstances, and drew up their own bye-laws for the regulation of the system enforceable by a court. In many areas the farmers collectively elected officers to manage the field system: the two most important were the foreman who would oversee farming activities, and the pinder, responsible for rounding up stray animals and locking them in the pound or pinfold.

This description of a commonfield system is inevitably a stereotype since no two villages were exactly the same. Field systems would be adapted to local circumstances and each would have its idiosyncrasies. Moreover most field systems were continuously changing, evolving according to local needs, but also to changes in landholding and property rights. Some had three fields and a three-course rotation, but villages could have more fields and longer courses, or manage to fit more courses into their three fields by adopting a unit of rotation other than the field. Thus sometimes the unit of rotation was not the field but the furlong, so that quite complex cropping sequences might exist despite there being three fields. The majority of commonfield systems were to be found in a band of country stretching from Dorset in the south-west to North Yorkshire in the north-east but they could also be found elsewhere. The research to produce even a moderately accurate map of commonfield systems has not been carried out, so it is only possible to give a very general indication of where they were. Figure 4.2(a) shows the distribution of commonfields *c*.1600 as published by Gonner in 1912. Another way of looking at the distribution of field systems is shown in Figure 2.1, which maps the settlement geography of England and Wales. This can be taken as a rough and ready guide to the former distribution of commonfield systems, if it is assumed that this field system was associated with nucleated settlement. In the Midlands there does appear to have been some uniformity in the topography, property rights and cultivation practices of many field systems, in that fields were subdivided, had common property rights, and were cultivated according to a three-field rotation with common grazing on the stubble and fallow. The topography of that system is shown in stylised form in Figure 2.2(a). The field system consists of strips and the holdings of a particular individual (in black) are evenly scattered.

However, this regular commonfield system was not ubiquitous. Table 2.3 reproduces a useful classification of arable field systems proposed by Campbell. It is based on the identification of fourteen key components of a field system. They can be grouped into those referring to topography (2–5), property rights (1, 6–10), and regulation (11–14). He describes the 'classic' Midland system as a *regular commonfield system*, characterised by

Table 2.3 *A functional classification of arable field systems*

	Non-common subdivided		Non-regulated cropping		Partially regulated cropping			FR	RCFS
			(i)	(ii)	(i)	(ii)	(iii)		
1. Communal use of waste	x	x	x	x	x	x	x	x	x
2. Closes and unenclosed strips	x	x	x	x	x	x	x	x	
3. Mainly unenclosed strips		x	x		x	x	x	x	x
4. Holdings irregular	x	x	x	x	x	x	x	x	x
5. Holdings regular							x	x	x
6. Full common of shack		x	x	x	x	x	x	x	x
7. Limited common on half-year fallow			x	x	x	x		x	
8. Limited common on full-year fallow				x	x	x			
9. Full common on half-year fallow				x		x	x	x	x
10. Full common on full-year fallow				x		x	x	x	x
11. Flexible cropping shifts					x	x	x	x	x
12. Regular rotation					x	x			
13. Manorial regulation								x	x
14. Communal regulation			x	x	x	x	x	x	x

FR Fully regulated cropping RCFS Regular commonfield system

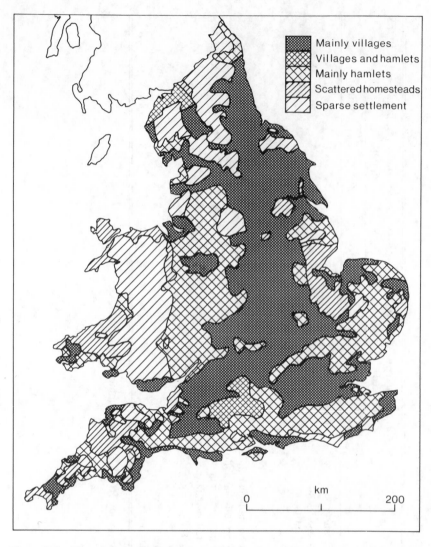

Figure 2.1 The settlement geography of England and Wales. *Source:* Thorpe (1964), 361–2.

regular layout, common rotations, and full rights of common grazing over the fallow strips. A variant of this system is *irregular commonfield systems with fully regulated cropping*, where holdings are composed of a combination of strips and small closes; a type that has been identified in woodland areas of the Midlands. *Irregular commonfield systems with partially regulated cropping* also had holdings consisting of a mixture of strips and closes (as in Figure 2.2(b) for example) and fallow strips subject to common grazing. However, unlike the two previous types, crop rotations are flexible and adaptable across the system. In some cases the fallow was grazed in common for a full year, in others from September to February. In East Anglia, for example, the unit of rotation was the 'shift' rather than the field or furlong. The flexible shift system could be easily adapted to vary the proportion of land sown with various crops and also enabled new crops to be introduced with relatively little difficulty. Rights to fallow grazing could vary. In some areas, full rights to grazing existed and the whole system was under communal control, as in the field systems of the Chilterns, parts of Essex and the Thames Valley. In other areas fallow grazing rights were limited, either by confining them to certain shifts so that crops could be grown on a 'fallow' shift, or by restricting rights to certain types of livestock or certain categories of farmer. The best example of this system is the East Anglian foldcourse of western Norfolk. A foldcourse represented the exclusive right to erect a sheep fold on the fallow and sometimes also over areas of permanent pasture. Ownership of foldcourses was a monopoly of manorial lords, and tenants had very limited rights to graze their animals on the fallow.

The fourth type of system, *irregular commonfield systems with non-regulated cropping*, was found in eastern Norfolk and in parts of south and east Devon. In these regions there were very few regulations constraining farmers' behaviour and most farmers were free to grow what they chose in the way they chose. Common grazing rights could still exist, as Table 2.3 shows; in Norfolk they were confined to the shack period after the harvest, and in Devon both to the shack period and to land lying fallow at other times of the year. Finally, there were some field systems *with non-common subdivided fields* where the common rights were confined to the waste, and, although the arable fields were subdivided, no common rights existed over them. Examples of this type include the subdivided fields of the Lincolnshire Fens, and, possibly, those of Kent.

Campbell's classification applies to arable field systems. *Infield-outfield* systems were quite common in areas where pasture was more abundant, particularly in parts of the north and west of the country where there was relatively little pressure of population on the land. As Figure 2.2(c) shows, the field system is divided into an infield and an outfield. The infield resembles an arable commonfield system and was cultivated along similar lines. Most of the outfield was under pasture, but from time to time parts were broken up for arable cultivation, and were then left to revert to grassland. As cropping in such systems became more intensive, the outfield could be enclosed as in Figure 2.2(d) and cultivated in severalty. Although these infield-outfield systems were found in the more sparsely populated areas and the classic Midland system in the more densely populated areas of the

(a)   (b)

(c)   (d)

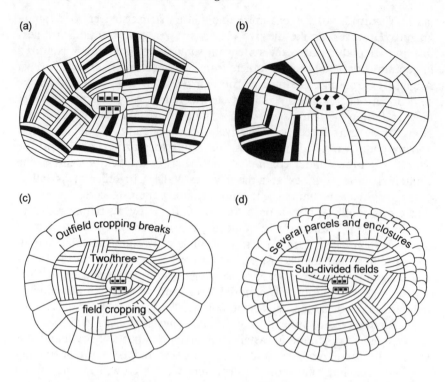

Figure 2.2   Types of field system. Based on Dodgshon (1980), 8, 48, 98 and 101.

country, there was no necessary connection between the type of field system and population density: some of the most densely settled parts of the country were to the south and east of the heartland of the Midland system.

Landholding

The common property rights discussed in the previous section are one aspect of the more general issue of landholding: the conditions under which land was held and was transferred from one person to another. The conditions of landholding are an important feature of social relations in the countryside, and, as with field systems, these conditions also imposed constraints upon farmers' freedom of action to farm as they wished. To understand the issue of landholding in the sixteenth century we have to examine the medieval legacy of both the theory and practice of land-holding under feudalism, since the situation in the sixteenth century is best understood in terms of its evolution, rather than by the retrospective imposition of modern categories. In the early middle ages, in principle all land belonged to the crown. Lords 'held' land directly from the king in return for services, and these lords in turn had people holding land from them in return for services. These services varied considerably, and were defined by the type of *tenure* under which the land was held. The duration of the tenancy, how long the land could be held, was defined by the system of *estates* which covered the rights to sell or lease the land to someone else, the rights to dispose of the land after death, and the rights applying to how the land could be used.

Some of the more common forms of feudal tenure are shown in Figure 2.3. In the early middle ages tenures in chivalry included knight tenure which required the tenant (a knight) to provide a certain number of horsemen to fight for the king. Spiritual tenures applied to land held by the church, and the services given in return were of a religious kind – saying mass, or giving alms to the poor for example. A third kind of tenure was socage tenure, which covered a variety of services including working on the lands of the superior lord. These three tenures were 'free' tenures, which meant that the services to be performed were fixed both in their nature and duration. With 'unfree' tenures (originally called villein tenure or base tenure), the nature of the service was not fixed and usually took the form of agricultural labour on the lord's lands. Originally a villein tenant had no legal right to land and in theory lords could demand of the villein what they chose, but during the middle ages the custom of particular manors defined the *de facto* rights that an unfree tenant had to his land. Thus the villein tenant held land 'at the will of the lord and according to the custom of the manor'. By the end of the fifteenth century, the royal courts were willing to assist in disputes over villein tenure, recognising the custom of the manor. When land under villein tenure was transferred from one person to another (say from father to son) the transaction was recorded in the manorial court roll. The tenant received a copy of this entry and villein tenure gradually changed its name to copyhold, which came to mean that land was held 'by custom of the manor and by copy of court roll'. Just as field systems varied considerably from place to place according to local custom, so did customary tenures.

In addition to services of a regular nature required in return for land, such as work on the lord's lands, there were various *incidents of tenure*, which were additional payments tenants had to make to their lord. For many freehold tenures these were often symbolic, a rose or a peppercorn for example, but for many copyholders with customary land they could be quite onerous. Customary land often had servile status associated with it. This meant that tenants, and their heirs, were liable for personal obligations to the manorial lord. A fine could be demanded when a tenancy was inherited (an entry fine), or sold to another farmer, when the tenant died (a heriot) or even when the lord's daughter married (a merchet). By the

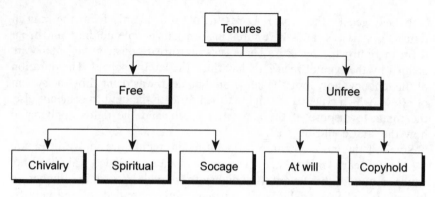

Figure 2.3   Tenures

sixteenth century most of these regular services and incidents of tenure had ceased to exist in their original form and virtually all direct labour and many payments in kind had been changed, or 'commuted' to a money payment. The incidents of free tenures in particular no longer carried much monetary value, so the land was held for practically nothing. But the incidents of tenure remained important because frequently they could be varied by the lord, whereas other services were fixed. Thus the payments made by a copyholder to his lord could comprise a low fixed yearly payment representing commuted labour services, but a much higher, and variable, entry fine or relief, when he took over the holding.

Nathaniel Kent, an eighteenth-century Norfolk land steward, described the situation as follows:

The copyhold is of two sorts, the one subject to, what is called here, an arbitrary fine, that is, a fine at the will of the lord, who, upon such estates, generally takes near two years value on descent, and a year and a half on alienation:– this copyhold is considered in value, about five years short of freehold. The other copyhold, is only subject to a fine certain, so that a lord of a manor can seldom take more than four shillings an acre, and sometimes only sixpence:– this is nearly of equal value to freehold.

It should be evident that it is misleading to interpret payments from tenant to lord as 'rent' in the modern sense since they may have borne no relation to the value of the land being farmed or to the profits of farming, and instead were the product of a long evolution of custom. Nevertheless, when custom enabled it, lords had a mechanism for matching payments to them to economic rents through the manipulation of entry fines.

By the sixteenth century the most common free tenure was socage tenure which had a secure title, was governed by common law and not by custom, and gave the tenant freedom to lease, sell and bequeath the land as he wished. The most common form of unfree tenure was copyhold. By the seventeenth century a new form of tenure, the 'beneficial lease', was granted for a period of years, or, more usually, for a life or period of lives.

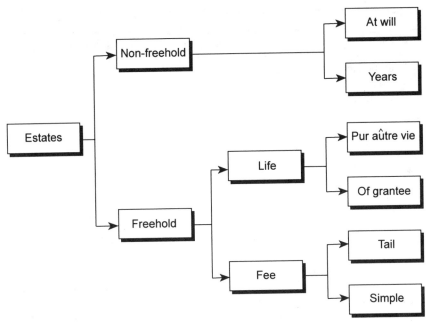

Figure 2.4   Estates

Rent payments consisted of a large initial payment, or fine, and regular small annual payments. Thus the holder of a beneficial lease was in a similar position to a copyholder for a period of lives.

The duration of a tenancy of land is termed the estate for which the tenant holds the land. The system of estates (Figure 2.4) cut across the system of tenures and was rooted in common law. The word 'fee' in descriptions of estates means that land held in this estate could be inherited. To all intents and purposes 'fee simple' gave totality of ownership, and from 1540, all land held in common socage tenure in fee simple could be devised in a will as the tenant chose. Land in 'fee tail' had restrictions imposed on the way it could be inherited: usually land could only be passed to direct descendants who would not have the right to dispose of it other than to their direct heirs. Another form of estate was for life. Land could be held just for the life of the tenant (of grantee), or for the life of the tenant, his wife, and his heir (pur aûtre vie). Thus the estate was bound to terminate at intervals and landlord and tenant had to renegotiate the estate. There were other forms of estate although they were much rarer. These estates were freehold estates, so-called because originally only someone holding a free tenure could hold one. By the sixteenth century the law had come to recognise the existence of estates in land held by copy-hold tenure although they were not called freehold. Thus copyholders who held land with non-free tenure could be *de facto* freeholders in the sense that their estate in land was for life or even in fee simple.

Table 2.4 *Regional variations in landholding (percentages)*

	Freehold	Customary	Leasehold	Uncertain
Northumberland and Lancashire				
11 manors	14	51	19	15
Staffs., Leics., Northants				
22 manors	18	62	14	5
Norfolk and Suffolk				
39 manors	36	54	5	4
South-west				
44 manors	20	61	13	7
Dorset				
49 manors	6	89	5	

Originally, the holding of land for a term of years did not constitute an estate; if someone gave land to another (a tenant) for a term of years the tenant held no rights under the law since he held neither a recognised tenure, nor a recognised estate. By 1500 a lease for a term of years had become recognised as an estate in land, but was classed as a non-freehold estate to distinguish it from the older forms of estate. Holding land by lease was also recognised as a form of tenure, since land was held in return for a service, in this case, money rent.

The legal theory of landholding is complicated enough, but in practice the situation was even more complicated. Evidence of how land was held is relatively common in early modern England, and is found in manorial surveys or 'extents', which list those holding land from a lord. Manorial surveyors usually divided tenants into three groups: freeholders, customary tenants and leaseholders. Freeholders were those holding land by free tenure, usually with the estate of fee simple, regulated by common law and outside the reach of manorial custom. By the sixteenth century most of their services were merely token payments which bore no relation to the value of the holding. Customary tenants were mostly copyholders, but there were also those holding land with no copy and therefore whose title was less secure. As we have seen, the position of the copyholder depended on local manorial custom. Some copyholders could transfer land as they wished and had fixed entry fines and heriots; on some other manors entry fines were arbitrary (although the law required them to be 'reasonable'), rents were subject to increase, and tenants had no right of nominating their successors. For example, the copyholders on the Bishop of Durham's estates at Wickham were exceptionally secure because by custom of the manor a copyhold was an estate in fee simple. This meant that their inheritance rights were guaranteed, and rents and fines were both small and fixed. Leasehold was a relatively new form of tenure, usually for an estate of years, but sometimes for a life or lives, and often applied to land that had been part of the demesne and thus not subject to custom.

Despite the survival of many thousands of manorial surveys it is difficult to arrive at figures of the proportion of landholders in these three groups. Many tenants held land in a variety of tenures: for example, in a study of East Anglia, of nearly 13,000 acres of land almost 60 per cent was held by tenants who were *both* copyholders and freeholders. Surveys also conceal the extent of sub-letting since lords were only interested in those who owed them some direct financial obligation and not in who was actually farming the land. Even so, given the abundance of sources, it is surprising that we still have to rely on figures shown in Table 2.4 collected by Tawney at the beginning of the present century. The figures in the table should be taken as a very rough guide. Although freeholders were quite prevalent in East Anglia for example, many held only small parcels of land, in contrast to the freeholders in the south-west who tended to own much larger acreages. As a very rough national generalisation it seems that in the early sixteenth century about three out of five tenants were customary tenants, more than one in eight were leaseholders, and about a quarter were freeholders.

In the south-west, copyhold tenure for three lives was by far the most common form of landholding, subject to the custom of each manor. An analysis of manorial rentals indicates that in Dorset almost 90 per cent of tenants were copyholders, but the conditions of copyhold tenure varied considerably from manor to manor so that in some cases copyholds gave almost all the advantages of freehold. In eastern England copyholds of inheritance were more common, whereby the land was often held as though it were in fee simple. Rules of descent varied however; often the land passed to the eldest son (the system known as primogeniture) but there were other forms of inheritance. Under the custom of borough English the land passed to the youngest son of the last wife, and under the system of gavelkind, descent was to all sons equally, or, in default, to all daughters. In the north of England the form of tenure was called tenant-right, which carried with it the obligation to ride out in arms against invaders from across the border. The gradual evolution of the laws governing land-holding, administered through courts ranging from those at the heart of the state in London to the remotest manor in the depths of the country-side, was responsible for the great variety and complexity of landholding in the English countryside in the sixteenth century. By the nineteenth century, this complexity had been reduced considerably; how that was achieved is discussed in Chapter Four.

Table 2.5 *Occupation and status labels of those engaged in farming recorded in Norfolk and Suffolk inventories, 1580–1740*

Craftsmen: Baker, Basketmaker, Blacksmith, Boller, Brazier, Bricklayer, Carpenter, Collar maker, Cooper, Cordwainer, Dornix weaver, Dyer, Fisherman, Glazier, Glover, Joiner, Leather dresser, Linen weaver, Mason, Mettleman, Millwright, Pail maker, Ploughwright, Plumber, Potter, Reedlayer, Rope maker, Rough mason, Shoemaker, Sievemaker, Tanner, Thatcher, Tiler, Timberman, Turner, Twisterer, Weaver, Wheelwright, Whitesmith, Woodsetter, Woolcomber, Worstead weaver

Specialist farming: Drover, Grazier, Marshman, Ploughman, Shepherd, Fisherman, Gardener

Professional: Apothecary, Chancellor at Law, Clothier, Doctor of Divinity, Merchant, Practitioner of Physic, Public Notary, Schoolmaster, Shipmaster, Surgeon

Retail/Service: Brewer, Butcher, Carrier, Chandler, Draper, Fellmonger, Fishmonger, Grocer, Innholder, Maltster, Mariner, Mercer, Miller, Oatmeal maker, Seaman, Tailor, Victualler, Waterman, Wool Chapman

Status: Alien, Baronet, Esquire, Gentleman, Gentlewoman, Husbandman, Knight, Singleman, Singlewoman, Spinster, Widow, Yeoman

---

Farmers and farm workers

The use of the word 'farmer' as an occupational label did not become current until the early eighteenth century. Before then, and indeed for some time afterwards, farmers were described with a variety of terms which denoted their status in the community rather than their occupation. In the sixteenth century, because most people were engaged in farming of one kind or another, the description 'farmer' by itself had little meaning and was not a label specific enough for differentiating people. Instead most farmers were classified in terms of their status rather than their occupation. Before examining this classification based on status it is necessary to set the occupation of 'farmer' into its early modern context.

Most employed people today earn their living by following a single occupation. In the sixteenth century probably the majority of people, including farmers, were usually involved in several 'occupations'. It is not surprising that the poor would turn their hand to anything if it brought in some money. Thus small farmers might take labouring jobs for other people, or they might have expertise in a particular craft or skill which they combined with farming. Some examples of these are given in Table 2.5, which is derived from some three thousand farmers in Norfolk and Suffolk with an extant probate inventory between 1580 and 1740 recording their status or occupation. They are defined as 'farmers' if their inventories record three or more items of farm crops or livestock. Inventories were only made for larger farmers, and rarely for farm labourers, so the evidence in the table is not a fair representation of all those working on the land, but it does demonstrate the range of occupations that were combined with farming.

The table shows how many higher-status and professional occupations were also combined with agriculture. Land was the most prevalent and the most secure form of investment in the early modern period, and while some landowners would have leased out their land, many retained a direct interest in farming on at least part of their holdings. Almost all clergymen were involved in farming, since the endowment for many livings was in the form of land (known as the glebe). Thus the Essex clergyman, Ralf Josslin, recorded in his diary for 3 April 1670: 'Cow calved; administered the sacrament, only 14 present.' Textile manufacture of some sort was the most common industrial activity combined with farming, although there were considerable regional variations. In the Pennines farming would be carried out in conjunction with lead mining, in certain forest areas with furniture making and basket weaving, and in others with the processing of hemp and flax to make rope and linen. Diaries occasionally give us a glimpse of the work patterns associated with this 'dual economy', such as that of Cornelius Ashworth of Wheatley in Yorkshire who in 1783, 'wove $2^3/_4$ yards the Cow having calved she required much attendance'.

People working on the land in the sixteenth century also differed from their modern counterparts in their attitude to work. The lives of all those working on the land were regulated by the rhythms of nature and the vagaries of the weather, but the seasonal agricultural cycle was also punctuated by religious and secular festivals. These included saints days and other holy days of the Christian calendar, but also secular events such as May Day. These festivals were more than simply a break from work since they involved communal activities which served to reinforce the bonds of community, especially in those open-field villages where farming was a communal activity. Examples of these ceremonies include 'plough Monday' in early January when the plough was ceremoniously carried round the village; and 'beating the bounds' held at Rogationtide, when parishioners would walk the boundaries of their parish. Many of these ceremonies were common across the country but individual communities would also have their own special customs and ceremonies. In the early sixteenth century there were fifty or more holy days a year (in addition to Sundays); how many of these were actual holidays is unclear, but an act of 1552 reduced them to twenty-seven. We have seen how work rhythms differed for arable and pasture farmers, but all farmers were subject to the limits of daylight and so worked longer hours in the summer than they did in the winter .

Underemployment caused by these interruptions, was, from the workers' point of view, involuntary, in that they had little choice as to whether they worked or not. But there was also much voluntary underemployment in early modern England, in that people chose leisure in preference to earning money. Economists represent this phenomenon by a labour supply curve that is backward sloping: as the price of labour goes up (the wage rate rises) the quantity of labour offered falls (people do less work). The explanation for this behaviour is that people worked for as long as it took to provide enough money for their basic needs: higher wages meant that these needs could be met with less work. This behaviour was probably a consequence of the absence of 'consumer goods' in the sixteenth century; there was little that extra income could be spent on.

The first four rows of Table 2.6 show how farmers were described by contemporaries according to their status. In ascending order, the farmers' status ladder rose from husbandman at the bottom, through yeoman, gentleman and esquire. Those ranked below husbandman may have been working on the land but would not have held enough land to be self-sufficient farmers; those ranked above esquire (in ascending order, knights, barons, earls and dukes) would have owned land but it is most unlikely they were actively engaged in farming themselves. Neither wealth nor birth were the sole determinants of status. Those in higher status groups were more wealthy on average than those of lower status, but the overlap between the groups was considerable so that it is impossible to predict farmers' status from their wealth. Table 2.6 demonstrates this using the valuations of goods and chattels (including livestock and crops) found in farmers' probate inventories in Norfolk and Suffolk. The status ladder is reflected by the ranking of mean wealth, but the standard deviation of the distribution of wealth for each group is very large. This indicates that the spread of wealth in each group is wide so that it would be impossible to determine rank simply on the basis of wealth alone.

Social origin was also important in determining status, but there was mobility between status groups, which might take time, perhaps a generation, but ultimately depended on wealth. Perhaps the most important determinant of status was the possession, and particularly the ownership, of land. The fundamental dividing line in the status hierarchy was between those with the rank of gentleman or higher and those below. For a farmer to be described as a 'gentleman' implied a great deal about his social and economic status. Most gentlemen owned the land they farmed; as a group they probably owned about a quarter to a third of the farmland of the country in the early sixteenth century (Table 4.8) although they constituted around 2 per cent of population (the proportion of gentry is higher in Table 2.6 because the inventories on which the table is based do not include those in the lower strata of society). They were usually educated, literate, and enjoyed a relatively high standard of material comfort in

Table 2.6 Wealth (£) and status of Norfolk and Suffolk farmers leaving probate inventories

	1584-1599				1628-1640				1650-1699				1700-1740			
	%	Mean	S.D.	Median	%	Mean	S.D.	Median	%	Mean	S.D.	Median	%	Mean	S.D.	Median
Esquire	2.3	380	159	405	1.5	697	465	517	2.5	778	537	580	0.5	328	251	328
Gentleman	7.0	205	164	165	8.0	289	248	192	13.2	569	633	408	8.0	441	489	291
Yeoman	38.1	140	122	99	43.0	250	269	155	42.4	249	257	188	53.8	255	242	176
Husbandman	27.9	63	70	42	15.1	83	69	59	7.0	89	89	59	8.9	155	174	90
Labourer					1.0	61	37	42	0.4	104	92	91	0.2	20	20	20
Farmer													1.5	282	196	293
Clergy	8.4	99	119	61	12.4	210	160	167	11.2	310	288	236	5.7	273	234	195
Widow	7.0	72	167	31	6.4	194	192	112	6.3	173	169	101	6.1	203	251	122
Retail/Service	2.6	66	75	26	2.1	356	357	207	4.1	226	202	169	4.8	261	205	172
Craftsman	6.5	51	62	34	9.6	129	129	85	11.6	166	184	97	10.0	176	179	116
Prof/Merchant	0.2	21	21	21	0.7	2454	1576	1881	1.3	1481	2252	550	0.6	125	77	142

S.D. Standard Deviation

households supported by servants. The gentry were acknowledged as such by their local communities and expected to carry out certain obligations: they would hold local public office, dispense their patronage, and be expected to offer hospitality to any of similar status. Most gentry also exercised lordship; they had jurisdiction, through a manorial court for example, over other people. Gentlemen were most unlikely to engage in the manual activities of farming. Thus in the 1590s Norden speaks of Middlesex farmers who 'wade in the weedes [clothes] of gentlemen; theis [these] only oversee their husbandrye, and give direction unto their servauntes, seldome, or not at all settinge their hand unto the plowgh ...'.

Below the rank of gentleman came the yeoman. Technically these were men who held land to the value of two pounds a year ('40 shilling freeholders') and this gave them political rights and a vote in parliamentary elections. But the term yeoman was applied to a much wider range of people. Some writers today use the term to refer to a vanished class of peasantry or self-sufficient family farmers for example. Marx applied it to his economic category of 'peasant', yet most yeomen worked large farms and employed labour as servants or day labourers, and, by the mid-seventeenth century, were more like proto-capitalist farmers than 'peasants'. Detailed studies show that in practice yeomen were differentiated from other groups by the size of their holding, whether it be owned, rented, or a mixture of the two. Yeomen were much more likely than gentlemen to do physical work themselves and were less educated than gentlemen (in the sixteenth century probably less than half were able to read and write). But they would play an important role in the life of their community and it was usually a yeoman who held the parish offices of churchwarden, overseer of the poor, and quarter sessions juryman. It is difficult to estimate the proportion of farmers described as yeomen because the figure varied regionally across the country. The inventory evidence in Table 2.6 shows that the proportion of yeomen farmers in Norfolk and Suffolk rose from 38 per cent to 54 per cent from the 1580s to the early eighteenth century: although the rising proportion may be representative, the size of the proportion is misleading because of the social bias in inventories. In 1522 the evidence of a muster which covered the county of Rutland showed that just over 2 per cent of the males over 16 years were called yeomen, but it is likely that the proportion of the population called yeomen rose considerably during the sixteenth and seventeenth centuries.

Husbandmen were the most numerous group of farmers: the Rutland muster evidence shows that over 55 per cent of those involved in farming were called husbandmen. They were on the lower rungs of the farming ladder and many were the descendants of medieval villeins. They were more likely to rent their land than to own it outright; much less likely to hold office in their local community; and were generally uneducated, with only 10 to 20 per cent able to sign their name in the late sixteenth century. Husbandmen were regarded as conservative and slow to change their ways: according to the seventeenth-century commentator, John Aubrey, husbandmen, 'go after the fashion, that is, when the fashion is almost out, they take it up'.

Another significant social divide was between those who could live from the land alone, and those who had to earn money in other ways to survive. Into this latter category fall cottagers, labourers and servants. Cottagers might have a small area of land attached to their cottage, but this would not be sufficient to support a family. Thus, like labourers and farm servants, they had to work for other people. The seventeenth-century statistician, Gregory King, reckoned that in 1688 over half the population could be described as 'Labouring people, out servants, cottagers and paupers' (764,000 families out of a total of 1,360,586). The proportion is likely to have been much smaller a century and a half earlier: the Rutland muster of 1522 describes 30 per cent of males as labourers and 9 per cent as servants. This is confirmed by a recent estimate that in the 1520s around 20–25 per cent of all adult males in southern England may have been labourers – landless men whose prime means of support for themselves and their families was by working for others.

The muster only counted males over the age of 16. Once boys, girls and women are added it has been estimated that between one third and one half of hired labour in early modern agriculture was supplied by servants in husbandry. Servants were unmarried men and women (in roughly equal proportions) living in the farmhouse as part of a farmer's family and hired on an annual contract. 'Labourers' were those hired for a shorter term and living off the farm. Servanthood provided a role for young people between puberty and marriage, and it has been estimated that 60 per cent of the population aged between 15 and 24 were farm servants. A sample of 100 parish listings records that 72 per cent of yeomen and 47 per cent of husbandmen had servants during the early modern period. Since servants lived in the farmhouse on an annual contract their labour was constantly available. Some women would be indoor servants, mainly engaged in cooking and cleaning, although women also ran the dairy, milked cows and cared for small animals such as pigs and poultry. Women also did outside work; picking stones, weeding crops, and helping with both the hay and corn harvests. Male servants would perform all the farming tasks, although on a large farm the particularly skilled tasks, such as managing horses, would be reserved for more senior workers. Some servants remained on a single farm until they married and became labourers or farmers themselves, but others moved around from farm to farm. In Northumberland married servants were hired on an annual contract and were provided with a house and specified payments in kind. The male servant (called a hind) undertook to provide female labour for the farm (his wife) for a cash wage.

There are very few surviving historical documents revealing the lives of servants and labourers – they were usually too poor to leave a probate inventory for example. A recent study from north Norfolk has revealed some of the complexities of the employment of servants and labourers by examining the account books of a gentleman's estate in the sixteenth century. For example, servants were not always hired at the same time of year, or for periods of a year; some servants were married and lived in accommodation provided by the farmer; and a sub-group of servants were effectively trainees or apprentices. The servants were the elite of the workforce on this Norfolk estate; they were entrusted with the skilled work with horses for example. Day labourers were divided into several groups. The specialists included shepherds, hedgers and ditchers, slaughterers, harness makers, molecatchers and carriers. This group received fairly regular employment, but the non-specialists were given relatively little work, which was mostly threshing, hay mowing, and general labouring duties. Finally, harvest labourers were itinerants just employed for the cereal harvest. The employment of labourers was irregular and most labouring families would have had an income insufficient for their subsistence from labouring alone, and were therefore engaged in other jobs within their community.

Contemporaries did not include 'peasant' as a category in their status hierarchy, yet historians have written a great deal about peasants and the peasantry. The word peasant has a general meaning of 'countryman', 'rustic', and more specifically as a small-scale subsistence producer working on the land, usually as an owner occupier. Sociologists and anthropologists have refined and extended these meanings, especially in a modern context, so that peasant societies are sometimes characterised as having extended households, communal property rights, and specific inheritance customs. Many of these definitions derive from studies in specific places and definitions are also influenced by ideologies ranging from the romantic to the Marxist. Much of the difficulty with the concept of the peasantry in early modern England stems from the fact that Marx identified a class of peasants in England that were swept away in the eighteenth and nineteenth centuries. Marx was concerned to account for the rise of capitalism, and his peasants were small farmers who survived independently of capitalist production. Those who disagree with Marx's characterisation of economy and society have therefore been keen to show that this was not the case; either by showing that such a class did not exist or was not swept away (this is discussed in Chapter Four); or by redefining the peasant so that the word no longer applies to the class that Marx was talking about. The existence or otherwise, therefore, of a peasant class is crucially dependent on definition.

Marx's definition raises the question of what we mean by capitalist production (to be discussed in Chapter Five), but his definition would imply that in the sixteenth century there were many peasants. Other definitions of peasantry imply the same. It is generally accepted that peasants were owner occupiers, who not only depended on the land for their living, but had strong emotional attachments to the land. However, they were not able to accumulate capital, so their holdings remained small and family incomes could be supplemented by wage labour. These peasants worked the land themselves and did not, as a rule, employ labour. It is difficult to generalise as to the size of their holding using this definition because it depends on the nature of the enterprise. Peasants also shared a common culture based on their common rights. Such a group would incorporate many of the husbandmen of the early sixteenth century, who had little or no capital, farmed mostly for their own needs, and were generally conservative.

However, at least one historian thinks a peasant class did not exist in the sixteenth century. Macfarlane constructs an 'ideal type' of peasantry, heavily based on east European experience, where the family was extended and multigenerational, land was held collectively, and its sale was unknown. English peasants had small households, were mobile, married late, and did not share the eastern European strong association with the land amounting to *de facto* family ownership and continuity of possession. Macfarlane considers, 'the majority of the ordinary people in England from at least the thirteenth century were rampant individualists, highly mobile both geographically and socially, economically "rational", market-orientated and acquisitive, ego-centred in kinship and social life'.

While it is a relatively straightforward matter to place individuals into particular categories within the social hierarchy of early modern England, and to describe the characteristics of the categories in terms of wealth, literacy, and so on, it is much more difficult to understand the relationships between the groups. These relationships are important, for they not only help us understand the lives of individuals in particular communities, but they also direct us towards the economic and social interactions between individuals which are essential to understanding the development of the agrarian economy. Wrightson has provided a useful way to analyse these by contrasting social relationships encouraging continuity and social identification, with those leading to change, or social differentiation.

Forces of identification included kinship, neighbourliness, and the relationship of paternalism and deference. Kinship, defined by blood or marriage, was more important with the higher status groups than lower down the social scale where neighbours were more important than all but immediate kinsfolk. Neighbourliness involved reciprocity, or give and take, between those who were effective equals. An example would be small farmers sharing equipment over and above the obligations that might be imposed by communal farming in certain areas of the country. Neighbourliness was also expressed in the intricate web of debt and credit that bound farming communities together. In contrast to neighbourliness, through which relationships were based on a balanced reciprocity of obligations, paternalism and deference (or patronage and clientage) were characterised by unequal obligations. Obligations were unequal because it was accepted that society contained permanent inequalities and therefore those with wealth and power were expected to help those less fortunate. Patronage was dispensed through charitable acts; helping the old, the sick, and farming tenants who had run into financial difficulties; finding employment for certain favoured clients, acting as security for a loan or mortgage, and so on. Those in receipt of paternalistic help were deferential and obedient in return and were therefore prepared to accept things as they were. Thus the relationship of patronage and clientage, like that of kinship and neighbourliness, served to maintain stability in rural society.

On the other hand, forces of differentiation were more likely to involve conflict and were potentially destabilising. These forces of differentiation were manifest in the relationships between employer and employee, and between landlord and tenant. Not all landlords were paternalistic and helpful to their tenants. The opposing interests of landlord and tenant ensured that on many occasions the relationship flared into conflict, which at its worst could involve protracted legal action or even violence. Given the complexities of landholding discussed above, such disputes were not uncommon, especially when a new landlord took over a manor or estate and interpreted custom in a rather different way from his tenants or from his predecessor. Examples of such disputes abound in court cases and reflect a gradual move towards an increasingly economic or contractual basis to relationships between landlord and tenant.

Another example of a source of tension was over the payment of tithe. In theory, farmers were supposed to pay one tenth of the value of the annual produce of their farm to the church. In practice payments varied because of a baffling combination of custom, case law and precedent. The right to the tithe was usually held by the local parish priest but it could also be held by a layman (called a 'lay impropriator') who acquired the tithe rights attached to land when it was sold by the church. Tithes were divided into great tithes (corn, hay and wood) and small tithes (wool, animals, animal products, and garden produce). Not all farmland was liable to a tithe payment, since a statute of 1549 exempted land reclaimed from the waste from tithes for the first seven years of cultivation. Tithes were often taken 'in kind', meaning that the parish priest literally took a tenth of all the agricultural produce of the parish; driving his cart into the harvest field and removing every tenth sheaf of corn, for example. Increasingly tithes were commuted to a money payment (called a modus). Records of the church courts list many cases involving disputes between the tithe holder and the farmers of a parish, which covered such issues as debt collection, whether certain crops were titheable, whether tithe was payable for crops grown on former waste land, and the falling real value of tithe payments made in cash as inflation gathered pace in the sixteenth century.

The local community

Social relationships were enacted within the local community. 'Community' is a word much used by historians yet its meaning is slippery. Sometimes it is concrete, referring to a particular place, but it is also applied to a more abstract concept of community, and 'community studies' refers to a method of analysis. Thus the concept of community is a complicated one but, to generalise, it embraces the importance of place or locality, a set of social networks, and shared institutions. Above all, the notion of a community implies belonging and sharing. As Hoskins puts it, 'Men lived in a place that had meaning and significance for them; their roots went down deep into the cultural humus formed by centuries of ancestors before them on that spot.' In the early sixteenth century, belonging meant being a participant in the life of a community and sharing not only institutions, but, in commonfield villages, work as well. With a population of only 2.3 million for the country as a whole the typical Midland nucleated village consisted of a community of thirty to forty households, and there were undoubtedly some villages where the idea of agricultural self-sufficiency extended to the self-sufficient community. To quote Hoskins again, 'in general people found all their earthly needs and wants met within a radius of three or four miles at the most, within sight of their own church spire'. Evocative as this picture is, it is an exaggeration, and untypical, if not for the Midlands then certainly for other parts of the country. From the sixteenth century onwards, long-distance migration became increasingly common, prompted by hunger and the need to survive (termed subsistence migration), and for employment or apprenticeship (termed betterment migration).

Migration was one factor that loosened community bonds, another was regional variations in settlement patterns and farming organisation. Community ties were at their strongest in the nucleated, commonfield settlements of Midland England, but in other parts of the country communities took very different forms. Bonds between neighbours were much looser when the community consisted of dispersed farmsteads farmed independently with few if any obligations imposed by common rights or communal cultivation practices.

Communities were defined in part by residential propinquity and social interaction but they were also formed by the working of local administrative units. The township, parish and manor were territorial units that conferred rights, required participation, and enforced obligations on those living within their boundaries. The smallest unit of territorial organisation was the township. It corresponded to a settlement and could also be the unit of agrarian organisation for control of the field system. Its inhabitants were therefore bound together by propinquity and communal agricultural practice. In much of southern England, parish and township were coincident, and it was rare to find more than one township in a parish. In the north, most parishes consisted of several townships. In 1811 (when the census enables a national picture to be drawn for the first time) the number of townships per parish ranged from more than seven in Northumberland to under one in the eastern counties from Sussex, through Kent, Essex, Suffolk and Norfolk. The parish was the local unit of ecclesiastical jurisdiction, for baptisms, marriages and burials, but it was also the focus of moral and ritual obligation. The parish also administered certain civil functions, especially poor relief: an act of 1536 put the onus of responsibility for the poor on the parish, and from 1572 the parish was responsible for collecting money to relieve the poor.

Cutting across the administrative geography of townships and parishes was the manor: in Hoskins' words, 'the legal shadow behind the reality of crops and buildings, cattle and sheep, ploughs and carts'. The manor was a unit of landholding, in that it consisted of an area of land held by a lord, but it was also an area for the administration and regulation of landholding and the organisation and conduct of agriculture. In some areas, particularly in Midland England, manor, township and village were coincident and centred on a nucleated village. In other parts of England this was not the case: in some east Norfolk hundreds, for example, there were four manors or more in each village; in other areas, particularly the north and west, a single manor might embrace several townships. Even where a village had one manor it was not necessarily the case that all members of the parish would come under the influence of the manor – manorial tenants could form a sub-group within the parish and thus live under a different legal system from their neighbours.

Agricultural regions

Variation in administrative geography is yet another example of regional variations in the sixteenth-century economy and society already noticed in passing, including farming systems and practices, settlement patterns, field systems, landholding, and social structures. This diversity can be demonstrated quite simply by describing regional and local variations, but without interpretation or analysis it is merely illustrative: analysing and understanding this diversity is much more problematic. Generalisations are based on a system of *agricultural regions*. This section of the chapter will outline the regional divisions employed in the fourth and fifth volumes of the Cambridge *Agrarian History* and more recent attempts to capture regional diversity through a system of English *pays*. However, identifying regions in the past is not a matter of discovery as is implied by the description of these regional structures, but of creation. There is no single set of absolute and invariable regions that existed in the past which historians merely have to discover: different regional frameworks can be constructed for different purposes. Inevitably therefore, a single set of regions is bound to be unsatisfactory as a universal description of regional variation.

The imposition of a single set of regions becomes much more problematic if it is used for regional *analysis* as opposed to regional *description*. Analysis of the relationships between the various elements of the rural economy and society may be seriously misrepresented if they are forced to share the same spatial framework. For example, if soil type, farming system, settlement pattern, landownership and rural industries are all incorporated into the same regional framework, the implication is that they are linked together in some way at a particular scale of analysis. In fact, each of these elements of the rural economy has its own regional geography, which may not be appropriate for studying the other elements. With these two problems in mind, the impossibility of finding a single set of regions for a universal description of the rural economy and society, and the danger of assuming relationships between elements of the rural economy because they share the same regional structure, we can examine the farming regions constructed by historians.

Most of these follow contemporary descriptions and are based on soil types. Hoskins characterises such regions as follows: 'a territory, large or small, in which the conditions of soil, topography and climate (and perhaps natural resources also) combine to produce sufficiently different characteristics of farming practice and of rural economy to mark it off from its neighbouring territories'. Soil types have been adopted as the basis for many local studies of early modern farming. They also underlie the national farming regions described in Volume IV of the Cambridge *Agrarian History*. The argument is that light soils encourage arable farming, while heavier soils, especially combined with high rainfall, are conducive to pasture farming. The most distinctive arable type is described as 'sheep-corn' farming, because it was prevalent on light lands which were dependent on sheep to maintain soil fertility for arable crops. Pasture farming was carried out in regions described as 'wood-pasture' or 'open pasture': the former being areas of relatively recently cleared woodland mainly under permanent pasture, while the latter were the permanent pastures of mountains and moorland.

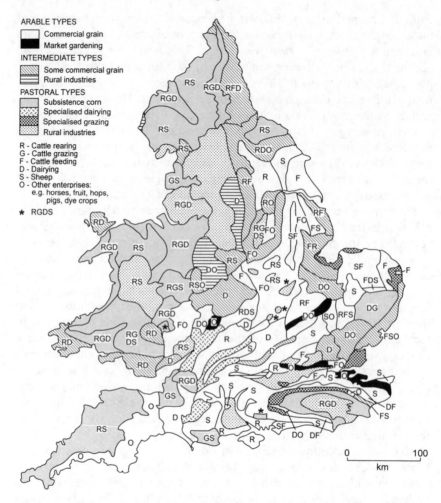

ARABLE TYPES
- Commercial grain
- Market gardening

INTERMEDIATE TYPES
- Some commercial grain
- Rural industries

PASTORAL TYPES
- Subsistence corn
- Specialised dairying
- Specialised grazing
- Rural industries

R - Cattle rearing
G - Cattle grazing
F - Cattle feeding
D - Dairying
S - Sheep
O - Other enterprises:
    e.g. horses, fruit, hops,
    pigs, dye crops

★ RGDS

Figure 2.5 English farming regions, 1640–1750. *Source:* After Thirsk. (1984b), xx.

The farming regions published eighteen years later in Volume V of the Cambridge *Agrarian History* (Figure 2.5) are more complicated than those found in Volume IV. There is more regional subdivision and the basis for division is changed slightly. Now the three types are categorised as 'arable', 'intermediate' and 'pastoral'. Regions are not defined simply on the basis of farming practice, but now explicitly include rural industries and the degree of commercialisation of agriculture. Farming is regarded as being more specialised in the period 1640–1750 compared with 1500–1640 but the apparent increase in complexity of the regional pattern in the latter period may well be the consequence of more detailed research. No doubt yet more

Table 2.7 *Sheep-corn and wood-pasture regions*

Characteristic	Sheep-corn	Wood-pasture
Land quality	Light	Heavy
Land availability	Shortage	Plentiful commons and wastes
Cash crops	Corn, wool	Dairy products, meat
Field system	Common, open	Several, enclosed
Settlement	Nucleated	Dispersed
Social control	Strong	Weak
Parish size	Small	Large
Population movements	Out-migration	In-migration
Industry	Little	Much
Social structure	Differentiated	Family farms
Politics	Conformist	Radical
Religion	Conformist	Dissenting
Crime	Order	Disorder
Sport	Team games	Individual games

research could add yet more regions to the pattern in Figure 2.5, but as the complexity of the regional structure increases it begins to negate one of the objectives of regionalisation which is to generalise about farming economies.

The linkage between agriculture and industry within these regional frameworks was identified from the outset. As Table 2.7 shows, the simple regional distinction between sheep-corn and wood-pasture has been adopted to show spatial variation in many other elements of the rural economy, society and culture. The sharing of the same regional framework by these elements suggests some necessary connections between them, and indeed historians have provided accounts of some of these connections. Thus sheep-corn regions were associated with subdivided fields and commonfield farming centred on old-settled nucleated villages within relatively small parishes. The more recently cleared wood-pasture areas tended to have enclosed fields farmed in severalty with dispersed settlement in larger parishes. Links are also made between the type of farming and the presence or absence of industry. Thirsk points out that the distribution of the wool textile industry was not coincident with the distribution of wool production. While spinning and weaving were more commonly found in wood-pasture areas, the raw material, wool, came from sheep in the sheep-corn regions of arable farming. The argument for explaining this spatial separation concerns labour supply. The labour demands of arable farming left little opportunity for farmers or their families to engage in some other occupation, whereas pasture farming was much less demanding of labour time and provided the opportunity for spinning or weaving in conjunction with farming. Dispersed settlement, private property rights and a fragmented manorial structure meant that social and economic control by a manorial lord was weak in wood-pasture areas, which further contributed to industrialisation since a lack of social control encouraged in-migration.

John Aubrey, the seventeenth-century antiquary thought there were far wider differences between the wood-pasture communities of north Wiltshire and the sheep-corn communities to the south of the county:

In North Wiltshire, and like the vale of Gloucestershire (a dirty clayey country) the Indigenae, or Aborigines, speake drawling; they are phlegmatique, skins pale and livid, slow and dull, heavy of spirit; hereabout is but little tillage or hard labour, they only milk the cowes and make cheese; they feed chiefly on milke meates, which cooles their braines too much, and hurts their inventions. These circumstances make them melancholy, contemplative, and malicious; by consequence whereof come more law suites out of North Wilts, at least double to the Southern parts. And by the same reason they are generally more apt to be fanatiques: their persons are generally plump and feggy: gallipot eies, and some black: but they are generally handsome enough ... On the downes, sc. the south part where 'tis all upon tillage, and where the shepherds labour hard, their flesh is hard, their bodies strong: being weary after hard labour, they have not leisure to read and contemplate of religion, but goe to bed to their rest, to rise betime the next morning to their labour.

While they have not yet gone as far as Aubrey (in regard to the physical features of the inhabitants for example) historians have tacked more and more onto the basic agricultural division between wood-pasture and sheep-corn. Some have incorporated it into their notions of 'protoindustrialisation': a model of industrial activity before the industrial revolution. Others have argued that nucleated settlement and strong manorial control in sheep-corn areas encouraged conventional and conformist attitudes to both politics and religion (see Table 2.7), while the absence of such social controls in wood-pasture regions meant that people living there were more likely to be radical and unorthodox in their beliefs. Thus it has been argued that wood-pasture cloth working areas were likely to have been more puritan in their allegiance, and also to have supported parliament in the civil war. Wood-pasture areas were also the locus of a higher incidence of crime, particularly anti-enclosure riots, than were sheep-corn regions. A further extension has used the distinction between sheep-corn and wood-pasture as the basis for a geography of sport in Wiltshire, with the cooperative farming regimes of sheep-corn areas promoting team-games, and the several farming of the wood-pasture regions promoting individualistic bat and ball games.

More recent contributions to the discussion of regional differences have introduced the concept of *pays*. This concept was developed by French geographers and is used to refer to distinctive countrysides that were a product of physical differences in geology, soil, topography and climate;

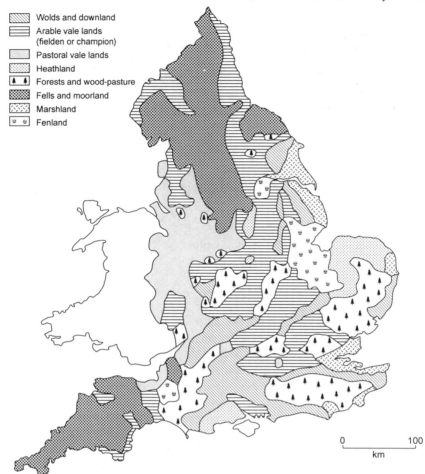

Wolds and downland
Arable vale lands
(fielden or champion)
Pastoral vale lands
Heathland
Forests and wood-pasture
Fells and moorland
Marshland
Fenland

0   100
km

Figure 2.6   English *pays* in the early modern period. *Source:* After Thirsk (1987), 39.

and also of differences in settlement history and rural settlement, which gave each *pays* a distinct way of life or *genre de vie*. Thirsk's map of English *pays* published in 1987 (Figure 2.6) is closely related to her original map of farming regions published in 1967, and is a considerable simplification of the map published in Volume V of the Cambridge *Agrarian History* (Figure 2.5).

The wolds and downlands in Figure 2.6 correspond to the sheep-corn areas and are characterised by arable husbandry supported by the sheep-fold. In the sixteenth century most villages were still under commonfields. Some heathlands, especially those in East Anglia, had a similar agrarian regime, but other heathlands, such as those in the New Forest, were very

Table 2.8 *Wealth distributions in 1522 (percentages)*

£	Arable farming		Pastoral vale		Forest		Highland	
	Persons	Wealth	Persons	Wealth	Persons	Wealth	Persons	Wealth
100 +	0.5	16.0	0.3	8.3	0.2	9.0	0.1	12.6
40–99	1.9	21.1	1.4	11.7	0.5	5.8	0.4	7.6
20–39	3.6	15.4	4.9	16.9	2.8	15.0	1.5	21.4
10–19	6.7	14.6	12.6	23.3	7.4	21.2	0.1	0.5
5–9	11.6	13.5	22.2	22.2	13.3	21.0	2.0	7.7
3–4	13.1	8.5	16.4	9.2	14.5	12.8	1.6	3.6
2	13.7	6.1	17.9	5.9	18.3	9.7	2.3	3.2
1<2	21.5	4.3	14.7	2.5	18.8	5.2	34.2	28.5
<1	27.5	0.5	9.6		24.2	0.4	57.8	15.2

Arable farming refers to Berkshire and Norfolk; Vale and Forest to
Gloucestershire, and Highland to Staincliffe in the West Riding of Yorkshire.

similar to the forest *pays*. Arable vale lands were also characterised by
commonfields and nucleated villages, and were lands of ancient settlement.
Depending on relative prices and market opportunities the land could
switch between grain and livestock: sometimes a move to livestock resulted
in enclosure. Pastoral vales had less nucleated settlement, and were more
likely to be enclosed. In contrast with the downlands, the social distribution of wealth was less extreme in these areas, as Table 2.8, based on taxation assessments, shows. The forest areas were not necessarily entirely
covered in trees. In the main they were late-settled regions, often with
extensive areas of scrub and improved grass and hence they are also called
wood-pasture areas. Settlement was dispersed, and forest areas were
attractive to squatters, resulting in a high proportion of poor people, also
evident in Table 2.8. In addition to pasture farming there were numerous
by-employments, often making use of forest resources, such as wood-
working and charcoal burning. Fell and moorland areas were late-settled,
gave rough pasture for hardy livestock, and also had relatively high
numbers of poor. The main grain crop was oats, and other activities often
included mining and quarrying. These impoverished areas, with little land
capable of producing corn and in remote areas of the country, were the
last regions to suffer crises of subsistence, since food had to be imported
to sustain the population, and supplies could not always be guaranteed.
The remaining two types of *pays* are the fenlands and marshlands.
Fenlands were regularly drowned by overflowing rivers. They were therefore mainly pastoral economies, supplemented by fishing and fowling, but
where arable land was available it was often very fertile. The fens were
inhospitable to outsiders, partly because of disease, and were typically

peasant communities. The marshlands were also primarily grazing areas, but had more arable than the fenlands and a more hierarchical social structure. They were also more accessible, less unhealthy and more fully exploited.

These *pays* provide the most useful way of generalising about regional diversity in early modern England, but they do have some limitations. They are appropriate for giving a general indication of the look of the landscape and the prevailing economic and social structures within an area, but they are more useful for descriptive as opposed to analytic purposes. By definition *pays* are homogeneous regions, tracts of country with a consistent character; as are the farming regions of Hoskins and Thirsk, in which a particular attribute, or set of attributes, is uniformly distributed. There are a number of reasons why the character of the rural economy may not in fact have conformed to this homogeneous model. Commercialised agriculture, which is a feature of the homogeneous regions in Figure 2.5, was probably not a characteristic of *all* the farms in a region, but confined to the larger ones. Where the natural environment was equally conducive to a variety of farm enterprises, as in the clay vales of Midland England, it might be expected that some farms might follow one activity and adjacent farms another, with no obvious pattern. In fact, closer empirical investigation, based on the measurement of farm types rather than impressionistic description, reveals that farming regions were not always homogeneous within a *pays* or farming region. Farm types are measured using probate inventories, which record the crops and stock on a farm when the farmer died. These data on crops and stock form the basis for a classification of farms, and Figure 2.7 shows the results of this exercise for some farms in Norfolk and Suffolk. The classification of these farms, using a technique called cluster analysis, results in thirteen farm types based on the proportion by value of the farm enterprise devoted to cattle, horses, sheep, winter corn, spring corn and fodder. The histograms in Figure 2.7 show the extent to which the proportions of each of these six characteristics deviates from the average for each of the thirteen farm types. Bars above the line indicate an above average concentration on the particular attribute, and bars below the line a below average concentration. The numbers above the symbols for each farm type are the number of farms falling into that particular cluster. Thus the first farming type, at the top left of the diagram, has below average proportions of cattle, horses and sheep; above average proportions of winter corn and spring corn; and an average proportion of fodder. The dendrogram at the bottom right shows how the thirteen types would merge if further iterations of the clustering process were to force a smaller number of clusters or farm types. Figure 2.8 maps these farm types using the symbols associated with each cluster in Figure 2.7 and shows the boundaries of farming types defined by Thirsk for the

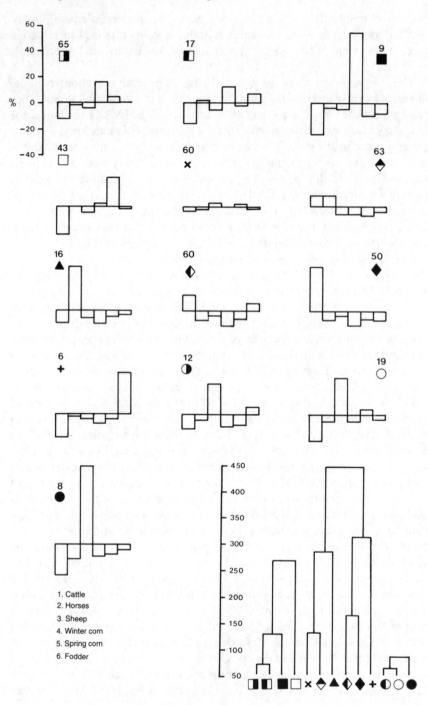

Figure 2.7   Farm types in Norfolk and Suffolk, 1584, 1587–96. *Source:* Overton (1983a), 16.

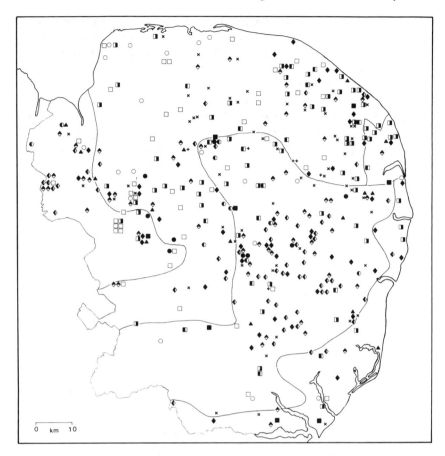

Figure 2.8 The distribution of farm types in Norfolk and Suffolk, 1584, 1587–96. *Source:* Overton (1983a), 17.

period in Volume IV of the Cambridge *Agrarian History*. The map clearly shows that in Norfolk and Suffolk, neither the *pays* in Figure 2.6 nor the farming regions in Figure 2.5 are homogeneous with respect to farm type defined in this way. A statistical measure of the extent to which farms of the same type are grouped together in space reveals that only farms of one type (the sixty-five farms in the first cluster defined in Figure 2.7) are concentrated together, and so drawing boundaries around farm types in Figure 2.8 would be inappropriate.

The categorisation of farm enterprises in this way is useful if the object of attention is the farm enterprise, in a study of the adoption of new crops by different kinds of farm for example, but other issues require different regional structures. The more appropriate construct for the study of marketing is the nodal or functional region, which is characterised by movement and interaction; by flows of goods and people. The marketing region, the area from which buyers and sellers travel to a market, is an example of such a region, and may overlap several uniform or homogeneous regions. Thus the provisioning zone of a city constitutes a nodal region which may be superimposed over many uniform regions based on farming type. One of the categories for the map of farming regions from the Cambridge *Agrarian History* in Figure 2.5 is 'commercial grain' which may more appropriately be defined in terms of a functional region rather than a uniform region since it implies interaction with markets.

However, the boundaries of such regions, defining the area provisioning a major town for example, are hard to draw. Within any area the larger farmers are much more likely to be involved in commercial production than are small farmers, and so no single boundary line could enclose all farms. The boundaries would also vary for different products: grain was more expensive to transport than, say, wool or cheese, and so the market area from which it came is likely to have been smaller. Livestock walked themselves to market and so could come from further away and were not tied to water transport, as was grain. The boundary of such a nodal region would also ebb and flow depending on the state of supply: in years of abundant harvest the range from which grain was supplied is likely to have been less than it was in years of dearth.

Thus while the general farming region or *pays* is useful for giving a general indication of differences across the countryside, more analytical examination of the rural economy and society demands more specific regions, appropriate to the particular question being addressed. The description of farming regions in Figures 2.7 and 2.8 is merely one example; other elements of the rural economy await more precise mapping. It is likely, however, that some will be fairly uniformly distributed (perhaps soil type and field systems), whereas others will not (perhaps landownership). This raises a particularly important question, namely the extent to which the elements of rural economy and society that are supposedly linked together within rural regions are in fact linked together in the way they are assumed to be. Discussion of rural regions has been conflated with discussion of the *relationships* between the elements of the rural economy described in Table 2.7. Some of the supposed relationships may be misspecified, when, for example, elements are incorrectly assumed to be uniformly distributed, or when they are linked together at inappropriate scales. The importance of these relationships extends beyond the question of regional diversity and gets to the heart of the workings of early modern economy and society. Where it is possible, the individual elements of the rural economy should be mapped separately, unless there is a demonstrable, consistent link between them.

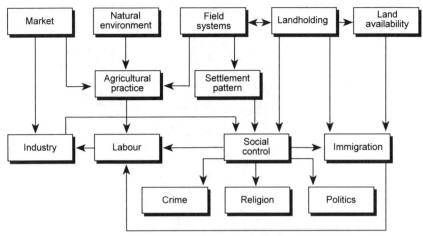

Figure 2.9 Some economic and social relationships in early modern England

Unfortunately very little research has attempted to measure the correspondence between the elements shown in Table 2.7, although it would be possible to investigate many of the associations quantitatively. Four of these will be discussed briefly here, albeit without the benefit of such investigations. They are shown in Figure 2.9 which is an attempt to show the relationships discussed in the literature of farming regions a little more clearly. The relationships to be examined are between the natural environment and agricultural practice; between agricultural practice and industrial activity; between agricultural practice and dissent; and between agricultural practice and individuality versus collectivism.

The assumption that agricultural regions should be based on soil regions is understandable, particularly for an era when agricultural technology (especially for draining land) was relatively unsophisticated, but soil type did not determine farming practice as many writers imply. More important, using predetermined regions based on soil introduces an inevitable circularity into explanations of regional differences. Agricultural practice is described within soil regions and the differences between regions is then explained in terms of differences in soil type. Many soils in England were (and are) equally suitable for a wide range of farming enterprises, even without technical or husbandry changes in farming systems. Figure 2.9 makes the point that agricultural practice is also the product of market influences and the local field system, and these influences could be supplemented by the force of local custom and tradition.

Soil conditions were nevertheless extremely important for early modern farmers and the soil regions shown in Figures 2.10 and 2.11, based on twentieth-century data, give some idea of the diversity of soils facing English farmers in terms of their ease of working and overall quality.

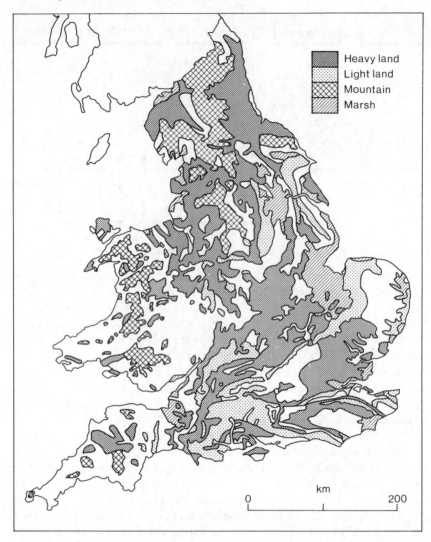

Figure 2.10 Heavyland, lightland, mountain and marsh in the sixteenth century.
Source: Avery, Findlay and Mackney (1974).

Figure 2.10 gives an indication of those areas of light soils and heavy soils, based on twentieth-century soil surveys (the areas of marsh are as they are likely to have been in the early sixteenth century). Light soils are easy to work, but dry out quickly and tend to lose their nutrients through leaching. In the sixteenth century they mostly supported sheep with rye and barley, although the lightest (in the Breckland of East Anglia for example) gave very low yields of grain. Heavier soils can be more fertile but are difficult

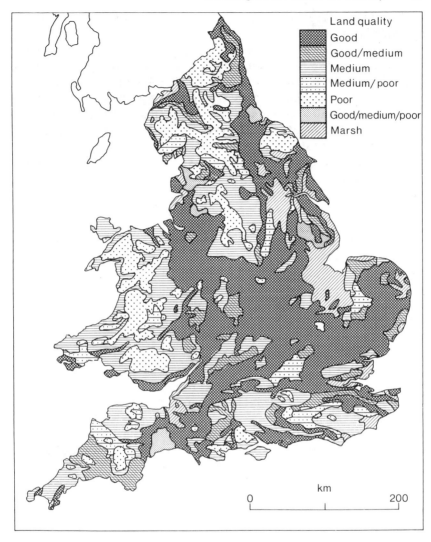

Figure 2.11 Land quality in the 1930s. *Source:* Stamp (1948), 383.

to work. Most are clays, which are difficult to plough, difficult to break down to a fine tilth, and prone to waterlogging. Figure 2.11 is a more sophisticated measure of soil quality, taking account of a wide variety of soil attributes influencing productivity, including depth, water conditions and texture; and also the characteristics of the site, including aspect. With the exception of marsh it refers to Britain in the 1930s not the 1530s, but even so gives some indication of conditions in the earlier period. The most important changes to soil conditions have been as a consequence of soil underdraining in the nineteenth century, especially in the heavy soil areas indicated in Figure 2.10.

Rural industrial activity has been associated with pasture farming because tending livestock requires less labour than growing crops. Those working on a family dairy farm, for example, would be occupied milking cows in the morning and evening, but could spend the rest of the day in handicraft production. It has also been argued that industrial activity was focussed on those areas where the poor quality of the land made it difficult to make a subsistence living from farming so that incomes had to be supplemented by alternative employment. While pasture farming was indeed linked to handicraft activity in some regions, there were also areas where industrial activity was associated with arable farming. Worsted weaving, for example, was concentrated in the villages to the north of Norwich, in an area of sheep-corn farming. Arable Bedfordshire had a thriving rural industry making pillow lace. The key to the distribution of rural industry that might be described as 'footloose' is not so much pasture farming as labour availability, thus Figure 2.9 intersperses labour supply as a category between agricultural practice and industry. Pasture farming was one reason why labour could be available but there were others. Many arable farming operations were the preserve of men and while women and children did undertake certain jobs on an arable farm, they could still have plenty of time for handicraft activity. Arable farming was more seasonal than pasture farming and links can be made between agriculture and industry on the basis of seasonal as opposed to daily fluctuations in work patterns. In the Sheffield region, for example, metal workers worked from March to August and then stopped for the harvest. Whatever the agricultural regime, an influx of immigrants, attracted by common land or by the absence of a large resident landlord could provide a pool of labour for handicraft activity. Inheritance customs could also affect labour supply: partible inheritance, for example, could lead to excessive fragmentation of holdings necessitating alternative employment. These factors linking agricultural practice, labour availability and industrial activity operate at different scales. The farm is the unit for the association of labour capacity with pastoral farming; the manor or the village is the unit over which some control over in-migration might have been possible; whereas the presence of an area of common, of forest, fen, heath or marsh could extend over an area embracing several villages with varied farming systems. Detailed local evidence of occupational distributions adds further credence to the view that a simple spatial correlation between pasture farming and industrial activity is too simplistic.

The link between the type of agriculture and dissent, in both religion and politics, is primarily through the mechanism of social control, which, as Figure 2.9 shows, is related to the extent of industrial activity, the settlement pattern, and the pattern of landholding. As with the control of in-migration, the unit for social control is the manor or parish and one of the main mechanisms of control was the presence of a resident lord. Social control was more difficult in areas of dispersed settlement. Dissent has also been linked to the time spent working on the land; pasture farming gave more time for reading and therefore for the encouragement of political and religious dissent. Even so, to argue that woodland or pasture areas are associated with dissent is only valid provided the connection is made through landholding structure and settlement pattern, which are not necessarily related to land use and farming practice. It seems that the presence of rural industry is, in itself, the key factor influencing local differences in levels of popular disorder and crime. Areas that were almost completely reliant on rural industry were more likely to be troublesome than those in which a dual-economy was in operation.

Landholding operated at the scale of the manor or parish, or occasion-ally, in the case of a large landowner, over several parishes. In the nine-teenth century contemporaries distinguished between 'open' and 'close' parishes. This did not refer to the field system, but to whether the parish was open or closed to settlement and in-migration. Open parishes were therefore more populous than close, and had a greater incidence of poverty, higher poor rates, and a labour surplus which could commute to work in adjacent close parishes. These close parishes, under the control of a single large landowner, had few opportunities for settlement, low poor rates and perhaps a deficiency of labour. This nineteenth-century model has not been explicitly applied to the sixteenth century, but some of its elements might well be relevant to the earlier period, particularly the domi-nating influence of a large landowner in shaping the economic and social development of the village community.

The final relationship to be considered is between agricultural practice and the prevalence of individuality or collectivism in a rural community. The supposed link here is between arable, commonfield, cooperative farming and collectivism on the one hand; and pasture farming, private property rights and individualism on the other. However, not all arable farming areas were areas of subdivided commonfields where farming was a collective activity. The sheep-corn region of the Chilterns, for example, was an area with dispersed settlement, irregular field systems and extensive commons. There were no open-fields in Kent although the county had plenty of arable husbandry. Some distinctive communities, such as the fenland for example, could be individualistic in day to day exploitation of resources yet could band together and act collectively against outsiders.

With the majority of farmers farming for subsistence and such regional diversity in local farming economies it might be argued that a 'national agrarian economy' could not exist in early modern England. It is impossible to speak of agriculture as a single enterprise, with a uniformity of objectives, but there are two ways in which we can speak of a national agricultural economy. First, the abstract idea of the agricultural economy still has a meaning when we say that agriculture was by far the most important contributor to GNP, in terms of national output and in terms of numbers employed. In other words, we can still apply the yardsticks by which economic sectors of the economy are evaluated today. In this sense farmers at opposite corners of the country who were not linked in any way at all to one another were still part of the agricultural sector of the English economy. Second, we can look at the extent to which the diverse farming economies were integrated together. It is extremely difficult to measure this: although most farmers were farming primarily at subsistence levels, they nevertheless could have quite extensive dealings with the market. But as we shall see in Chapter Four, the 'market' was their local market, and it is likely that there was little integration of these local market areas.

Conclusion

The picture of the rural economy described in this chapter is one of variety. The countryside was infinitely more varied than it is today, with respect to farming systems, landscapes, field systems, landholding, social structures and cultural experiences. The majority of farmers were producing to service their own immediate needs, yet all had dealings with the market and a few were farming on a large scale that was to become commonplace three centuries later. Geographical variation in each of these elements of the rural economy is easy to demonstrate, and the *pays* defined by Thirsk are a good general guide to regional variation in the face of the countryside. However, their role is better suited to description rather than analysis. In fact, assumptions about the geographical co-variation in the elements of the rural economy have, at times, been misleading, and it is sometimes more helpful to forget about geography when exploring the relationships between the elements of the rural economy shown in Figure 2.9. Some developments over the three and a half centuries from 1500 served to reduce the complexity and variety of the rural economy, but it is also the case that variety and diversity were increased by the introduction of new crops and farming systems. The transformation of farming techniques is the concern of the next chapter, while Chapter Four deals with the equally important transformation in social structure.

3

Agricultural output and productivity, 1500–1850

Included in Gregory King's many calculations made during the final decade of the seventeenth century are his estimates of the future population of England. He expected the population of the country to grow from 5.5 million in 1700, to 6.42 million by 1800, and to 7.35 million by 1900. In fact, as Figure 3.1 and Table 3.5(a) indicate, the population of England stood at 8.66 million in 1801 and 30 million in 1901. His forecast for the maximum population of the country was around 11 million people, which he did not expect to be reached until the year 3500 (in fact it was achieved by 1820). King's projections were so wide of the mark because he thought the country had insufficient land to support more people. His assumption was that, with a finite area of land and an agricultural technology that was virtually static, anything other than a very gradual growth in population was impossible. This assumption was based on a lesson from history. Before the mid-eighteenth century English population seemed to have a natural ceiling of around 5.5 million people. Whenever population grew (during the Roman occupation, in the thirteenth century, and again in the sixteenth century) agriculture had the greatest difficulty in meeting the increased demand for food. In each case there appears to have been a check: the rise in population was halted because the increase in agricultural output was insufficient to sustain the rise in population.

This chapter is concerned with why King's figures were incorrect; in other words with the transformation in output that enabled unprecedented population growth. It will become apparent that agricultural output can be increased in many ways (not just by the mechanism outlined in Chapter One), but that one of the most significant ways is through an increase in land productivity. The chapter starts by charting the course of population and agricultural prices, demonstrating how trends in the two were inextricably linked until the end of the eighteenth century. The attempt is then made to provide a chronology of trends in output and productivity, and the bulk of the chapter is concerned with the processes by which agricultural output and productivity rose.

Table 3.1 *Wheat, barley, oat, beef, mutton and wool prices, and agricultural wages, 1500–1849 (10 year averages 1700–49=100)*

	Wheat	Barley	Oats	Mutton	Beef	Wool	Wage	Real wage
1500s	22	17	15	–	–	24	32	146
1510s	23	18	17	–	–	31	32	141
1520s	29	22	21	–	–	28	33	116
1530s	28	25	22	–	–	31	35	124
1540s	34	34	27	–	–	39	37	109
1550s	57	64	50	–	–	53	51	89
1560s	58	53	46	–	34	53	56	96
1570s	67	59	48	–	34	60	66	97
1580s	77	76	65	–	37	58	64	83
1590s	100	94	90	–	47	81	70	69
1600s	95	92	85	82	51	89	70	72
1610s	112	105	102	89	58	91	72	64
1620s	113	102	89	91	57	91	80	71
1630s	133	138	112	95	62	105	91	68
1640s	143	125	119	103	72	100	96	67
1650s	117	99	112	121	75	129	96	82
1660s	114	98	103	125	76	134	96	84
1670s	112	94	99	116	76	122	96	86
1680s	96	91	100	119	76	103	98	101
1690s	131	104	108	114	74	127	99	75
1700s	100	95	94	100	76	113	97	97
1710s	107	104	104	109	102	107	99	92
1720s	108	111	108	104	105	94	97	89
1730s	93	97	98	90	106	91	102	108
1740s	91	94	97	96	112	95	105	114
1750s	109	121	106	100	106	96	–	–
1760s	120	139	118	112	123	100	–	–
1770s	147	138	133	134	153	100	110	74
1780s	146	132	137	140	152	113	129	88
1790s	182	173	181	183	205	166	143	78
1800s	267	240	263	259	303	237	209	78
1810s	288	260	279	266	295	267	209	72
1820s	189	182	213	214	265	168	183	96
1830s	179	181	204	215	261	210	172	95
1840s	176	182	179	218	268	172	162	91

Population and prices

Figure 3.1 and Table 3.5(a) show the results of many years of labour by the Cambridge Group for the History of Population and Social Structure. While their figures of English population totals may be subject to a small

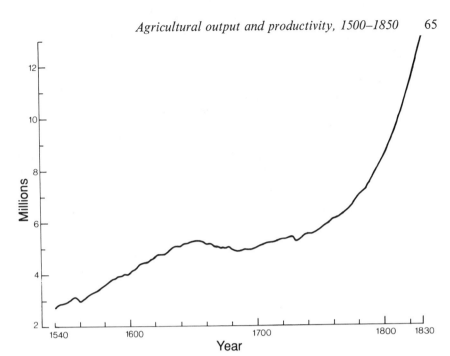

Figure 3.1 English population, 1541–1831. *Source:* Wrigley and Schofield (1981), 531–5.

margin of error they are substantially correct, and reveal a straightforward trend. English population rose from under 3 million at the start of the sixteenth century to 5.3 million by the mid-seventeenth century. Thereafter population fell slightly, but recovered its former peak during the second decade of the eighteenth century. From the 1730s population began to grow at an unprecedented rate, and that rate of increase was sustained into the nineteenth century, with the population reaching 8.7 million by the turn of the century and 16.7 million by 1851.

The long-term trends in prices shown in Figures 3.2 and 3.3 and in Table 3.1 are in part a consequence of changes in demand which are directly related to population change. Generally speaking, when population rose prices also rose. Taking wheat as an example, prices doubled from a ten-year average index of 22 in 1500 to 57 by 1550, and had almost doubled again to 95 by 1600. The rise continued into the new century, reaching a peak of 147 in 1642. For the next 120 or so years the index remained in the range 92–133, but then it began to rise again, as population rose, reaching a high of 296 in 1809 (the yearly peak as opposed to the ten-year average was 399 in 1812). From this peak prices start to fall, despite the continued rise in population. The price of meat (beef and mutton in Figure 3.3) is more difficult to document and the series may be less reliable, but also exhibits similar trends.

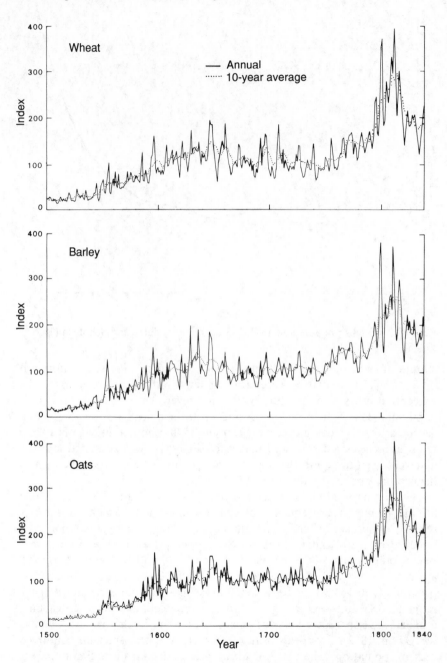

Figure 3.2 The prices of wheat, barley and oats in England, 1500–1840 (annual figures and 10-year means expressed as index numbers, 1700–49 = 100).
Source: See Table 3.1.

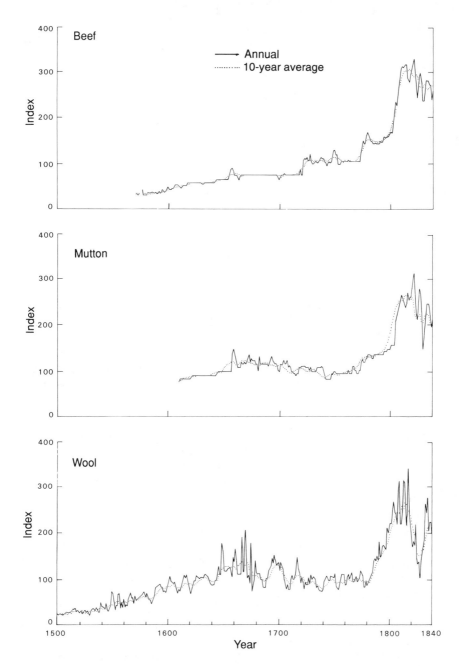

Figure 3.3 Prices of beef, mutton and wool in England, 1500–1840 (annual figures and 10-year means expressed as index numbers, 1700–49 = 100).
Source: See Table 3.1.

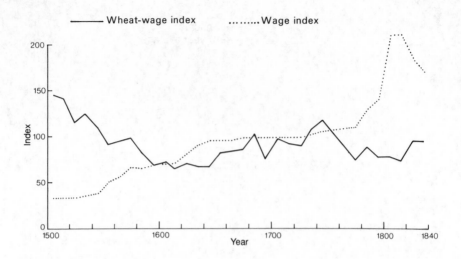

Figure 3.4   Agricultural wages in England, 1500–1840 (real wages defined as agricultural day wages divided by the price of wheat; 10-year averages of pence per day expressed as index numbers, 1700–49 = 100). *Source:* See Table 3.1.

Although the correlation between the movement of population and prices is a close one, population growth was not the sole determinant of the level of prices. In the short term, fluctuations were the consequence of changes in supply as a result of the harvest. Over the longer term the import of precious metals from the new world into Europe increased the money supply in the sixteenth century, and was undoubtedly responsible for raising the price level, as were the debasements of the coinage in the 1540s. But these were only partly responsible for the general rise in the price level: population growth was the major factor. The impact of increased demand through population growth is evident in the differential rates of increase for agricultural and industrial goods, and, within the agricultural sector, for cereals and meat. Increasing consumer demand put more pressure on cereals because they were the cheapest form of protein and so prices rose more rapidly than those for meat. Wages rose slightly during these inflationary periods but they lagged behind prices. Thus when wages are expressed in real terms (that is in terms of what they will actually purchase) they show an inverse relationship to prices. The real agricultural wages in Figure 3.4 and Table 3.1 are expressed in terms of the amount of wheat that the average wage will purchase (the wheat-wage index) and they show a fall from 1500 to the early seventeenth century, a hesitant rise to the 1740s, but thereafter a continued fall as prices rise once again.

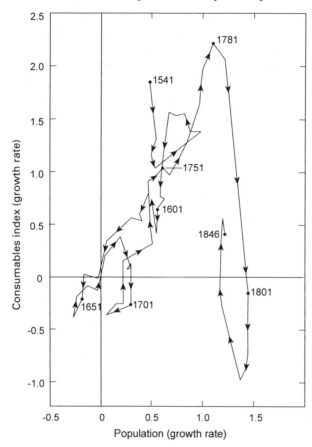

Figure 3.5 The relationship between the rate of growth in prices and the rate of growth in population in England, 1541–1846. *Source:* Wrigley and Schofield (1981), 405.

The relationship between the *rate of growth* in population and the *rate of growth* in prices is shown in Figure 3.5. It indicates a strong positive relationship from the 1540s to the 1780s: the rate at which prices were growing followed the rate of population growth. But after the 25-year period starting in 1781 the relationship changes: population growth rates rise to unprecedented levels (over 1 per cent per annum), but the rate of growth in prices starts to fall, from a peak of over 2 per cent per annum. This change in the relationship suggests that an important change had taken place in agricultural supply. Although population was growing, the agricultural sector of the economy was able to expand output to meet the additional demand, so that prices failed to rise so rapidly as they had done under pressure of demand in previous centuries. Figure 3.5 makes the same point as the discussion of Gregory King's population estimates: at the start of the eighteenth century English agriculture seemed unable to expand output significantly, but by the end of the century such expansion was well under way. This conclusion is based on the opinion of a famous contemporary, and the indirect evidence of prices and population: to go further it is necessary to provide some quantitative estimates of agricultural output and productivity.

Estimates of agricultural output and productivity

Comprehensive, nation-wide agricultural statistics were not collected in England until 1866. Thus one of the main problems facing historians of agricultural development is simply charting the course of agricultural output. In the absence of such output statistics, some historians have inferred changes in output from changes in farming practice. This has sometimes led to circular arguments, in which output changes are inferred from changes in cropping, so that the appearance of new crops like turnips and clover (which are fairly easy to document) is assumed to lead to changes in land productivity (which is difficult to document) and thus held responsible for changes in output. A circular argument is not necessarily an incorrect one, and the appearance of turnips and clover could well have led to changes in the output per sown acre of cereal crops. However, for their full benefits to be achieved they had to be cultivated in certain ways in conjunction with other developments in farming. But, as we shall see, these two crops represent only one of many possible means towards increased output, and that increase in output could be brought about by a wide range of agricultural changes, some involving productivity change and some not.

As a prelude to discussing these changes it is necessary to clarify the distinction between output and productivity. Agricultural *output* simply refers to the products produced by farmers; since some of these products become agricultural inputs (as seed or as food for livestock) output is best considered as the output available for human consumption. 'Productivity' is a much abused word, but it can be simply defined as the ratio of output to input over a given period. In practice, productivity indices vary considerably, depending on the combinations of outputs and inputs that are considered and the units in which they are measured. Productivity can be measured by relating *physical* quantities, which is appropriate for studying technological change and will be adopted here, but it can also be measured by *economic* indicators of the value of inputs and outputs. In an historical context limitations of available data necessitate using a variety of measures of productivity, particularly of land productivity, which can be confusing if their relationships are not clarified.

For these reasons the relationships between agricultural output, and

Table 3.2 *Output and land productivity*

Output:	
Q	= QCH + QA
QC	= TC × YC
QCH	= QC – QS – QCF
QA	= A x YA
A	∝ QFOD
QFOD	= QG + QAF + QCF
QG	= TG × YG
QAF	= TAF × YAF

Area:	
TAR	= TC + TAF + TFAL
TFOD	= TG + TAF + TFAL
T	= TAR + TG

Land Productivity:	
Q / T	= Gross land productivity
QC / TC	= YC – Gross cereal yield per unit sown area
QCH / TC	= Net cereal yield per unit sown area
QC / TAR	= Gross cereal yield per unit arable area
QCH / TAR	= Net cereal yield per unit arable area
QA / A	= YA – Output per animal
QA / QFOD	= Animal output per unit of fodder
QA / TFOD	= Animal output per unit fodder area
QFOD / TFOD	= Fodder output per unit area
QG / TG	= YG – Fodder yield outside arable rotations
QAF / TAF	= YAF – Fodder yield in arable rotations

Key:

A	Number of animals
Q	Output
QA	Output of animal products
QAF	Output of fodder crops in arable rotations
QC	Cereal crop output
QCF	Cereal crops fed to animals
QCH	Cereal crops for human consumption
QFOD	Output of fodder crops
QG	Output of grass outside arable rotations
QS	Seed for following harvest
T	Total agricultural area
TAR	Total arable area
TC	Cereal crop area
TAF	Area of fodder crops in arable rotations
TFAL	Area of fallow
TFOD	Area of fodder crops
TG	Area of grass outside arable rotations
YA	Output per animal
YAF	Fodder yield in arable rotations
YC	Gross cereal yield per unit sown area
YG	Fodder yield outside arable rotations

Table 3.3 *Yield per sown acre compared with yield per unit of arable area*

Rotation	WFFF	WWWF
TAR (acres)	40	40
TC (acres)	10	30
QC (bushels)	200	300
QC/TC (bushels per acre)	20	10
QC/TAR (bushels per acre)	5	7.5

W is wheat, F is fallow; see Table 3.2 for other abbreviations.

various measures of land productivity are formalised in Table 3.2. Agricultural output (Q) is defined as the output of cereal crops for human consumption (QCH) plus the output of animal products (QA). Cereal crop output (QC) consists of the area under cereals (TC) multiplied by the yield per unit sown (YC). The quantity available for human consumption is less than this because seed (QS) must be retained for the following year's crops and some cereals are fed to animals (QCF). Given the state of agricultural technology the total arable area (TAR) must also include fodder crops (TAF, such as peas, beans, turnips and clover), and an area of fallow (TFAL). The output of animal products (QA) is given by the number of animals (A) multiplied by the output of animal products (meat, dairy products, tallow, wool, hides and skins) per beast (YA). This is partly a function of how much they are fed, but it also depends on the rate at which animals convert their food into these products. Thus, for example, the output of meat depends on the number of animals, the quantity of food they receive, the rate at which they convert this food into meat, and how long they take to 'finish' or reach maturity and be ready for the butcher. The number of animals is proportional to the food available for them (QFOD) composed of grass and hay (QG) fodder in arable rotations (QAF), and cereals fed to livestock (QCF).

It should be evident from the equations in Table 3.2 that there are many possible ways of relating agricultural outputs to land inputs and so measuring land productivity. The most useful relationship, and the one most commonly used in modern studies, is simply the total output of agricultural products divided by the agricultural area (Q/T). In the absence of the data to calculate this, the index of land productivity more commonly used by historians of early modern England relates the output of a particular crop (usually wheat) to the area on which it was grown (QC/TC); this is usually calculated in terms of the volume of grain per unit of land sown, as bushels per acre. In early modern England sources are not available to measure productivity in terms of yield per seed (or output per unit of seed sown) although this index is used to measure medieval yields in England

Table 3.4 *The difference between gross and net wheat yields*

Bushels per acre			Net as % of gross	Growth rate (%)	
Gross	Seed	Net		Gross	Net
8	2.5	5.5	69		
12	2.5	9.5	79	50	73
16	2.5	13.5	84	33	42
20	2.5	17.5	88	25	30
24	2.5	21.5	90	20	23
28	2.5	25.5	91	17	19

and yields in parts of Europe. Yield per sown acre is the most commonly available measure for early modern England, was used by farmers and contemporary commentators, and is a good indicator of the performance of a particular harvest. However, yields calculated in this way may not be a reliable guide either to the overall course of land productivity or to output. Yields per sown acre are directly related to the area of fallow and fodder, so a farm with a relatively large fallow area might have higher yields per sown cereal acre (QC/TC) but much lower yields per unit of arable (QC/TAR) than a comparable farm with a lower fallow proportion. This is illustrated using hypothetical data in Table 3.3 from a farm whose arable is composed of just wheat and fallow. Yields per sown acre with a long fallow in the rotation (WFFF) are double those of a short fallow rotation (WWWF) although output per arable acre under the latter is 50 per cent higher than the former. It is also important to note that yields can be measured as gross or net of seed. Table 3.4 shows the amount of grain available for consumption after deducting the following year's seed requirements. When gross yields are relatively low, as they were in the early sixteenth century, seed takes a much higher proportion of the harvest than it does as yields rise. The phenomenon is more exaggerated for barley and oats because their seeding rates were higher, typically around four bushels per acre.

Livestock productivity can be measured in two basic ways; the first relating outputs from animals to inputs of fodder; and the second the number of animals to the area of land that can support them, in other words, livestock densities. Output per animal (QA/A) is partly a function of fodder inputs, but also reflects the efficiency with which animals convert food into saleable products (QA/QFOD). Livestock densities (QA/T or QA/TFOD) depend on the amount of animal fodder produced per unit area, or, in other words, yields of fodder (QFOD/TFOD, QG/TG and QAF/TAF).

Just as land productivity is calculated by dividing output by the land area input (Q/T) so labour productivity can be calculated by dividing output by the number of agricultural workers in the population. However, this measure of labour input does not account for the length of time those employed in agriculture actually spend working, so a complementary measure of labour productivity is derived by dividing agricultural output by the number of worker-hours per annum. The calculation can be further refined to take account of the respective contributions of men, women, children, and seasonal and part-time workers.

The measurement of output and productivity

Despite the absence of national statistics it is important to attempt to establish the levels and trends of output and productivity in the centuries before 1850. There are many ways of doing this, but they can be divided into 'bottom-up' and 'top-down' methods, and into direct and indirect methods. 'Bottom-up' methods are those that aggregate from observations made at farm or village level. On a farm scale, for example, some estimates of output can be calculated from farm accounts, but these are very scarce before the mid-eighteenth century. For a village, it is sometimes possible to reconstruct output from tithe records. Those owning the right to the tithe in a village sometimes collected agricultural statistics from the farmers in their parish in order to calculate the tithe payments to which they were entitled. A handful of these tithe accounts exist, but few have been analysed, and in any case very few survive for long periods of time.

'Top-down' estimates of agricultural output before the advent of nineteenth-century statistics can be made in three ways: by using the size of the population as an indirect indicator of the amount of food consumed; through the use of equations which in effect specify the demand curve for agricultural products; and by making direct estimates of the volume of output based on contemporary opinions. Table 3.5(a) shows the growth of English population from the early sixteenth century. Population has already been taken as a rough indicator of demand in the introduction to this chapter, but at least two assumptions must hold if the growth in population is to reflect a growth in agricultural output. The first is that no food was exported or imported. This is obviously incorrect, especially for the later period, so Table 3.5(a) also gives some rough estimates of net imports as a percentage of total output. The second assumption is that consumption per head of agricultural products (including industrial raw materials produced by agriculture) was constant. Given the trends of wages and prices, and the growing use of agricultural products in industry, this is extremely unlikely and so the estimates in Table 3.5(a) will tend to smooth out fluctuations.

The use of demand equations overcomes the second assumption inherent in the population method. Crafts has pointed out that output trends based

Table 3.5 *Estimates of English agricultural output, 1520–1851*

(a) Population method

Date	Population (millions)	Net imports (%)	'Output' index (1700 = 100)
1520	2.40	0	47
1551	3.01	0	58
1601	4.11	0	80
1651	5.23	0	101
1661	5.14	−1	100
1701	5.06	−2	100
1741	5.58	−5	114
1751	5.77	−8	121
1761	6.15	−4	124
1781	7.04	0	136
1791	7.74	+2	147
1801	8.66	+5	159
1831	13.28	+12	226
1851	16.74	+16	272

(b) Crafts' 'demand equation' method

	Index (1700 = 100)
1700	100
1760	143
1780	147
1801	172
1831	244

(c) 'Volume method' (value of total agricultural output in £m at 1850 prices)

	Value				Index (1700 = 100)
	Crops	Meat	Dairy	Total	
1700	17.00	12.87	6.05	40.12	100
1750	21.85	15.90	10.82	51.11	127
1800	31.99	21.36	14.78	76.46	191
1850	51.50	32.50	19.34	114.46	285

on population are inconsistent with the behaviour of agricultural prices. When agricultural prices are falling it is likely that *per capita* consumption will increase (assuming that people will consume more food because it is cheaper), and conversely when prices are rising *per capita* consumption should decrease. He therefore calculates output by taking prices and wages

Table 3.6 *Estimates of land use in England and Wales, c. 1700 – 1871 (million acres)*

	c. 1700	*c.* 1800	*c.* 1850	1871
Arable	9.0	11.5	15.3	14.9
Sown arable	7.2	9.7	14.3	14.4
Fallow	1.8	1.8	1.0	0.5
Meadow and pasture	12.0	17.6	12.4	11.4
Total	21.0	29.0	27.7	26.3

into account together with assumptions about the income and price elasticities of demand to produce the estimates which are shown in Table 3.5(b). While this technique overcomes a limiting assumption of the population method it introduces new assumptions of its own, since it is based on an idealised model of the economy which was unlikely to have existed in practice.

Direct estimates of the volume of output involve none of these assumptions, but are based on the estimates of contemporaries. Since agricultural statistics were not collected these can be no more than guesses, yet with the benefit of hindsight, they are surprisingly plausible. Gregory King, for example, thought England and Wales amounted to 39 million acres, which is an overestimate of only 1.675 million acres or 4 per cent. In some cases contemporary estimates have been revised and modern guesses have been used to interpolate the gaps, although these revisions are often based on the evidence of population growth and assumptions about *per capita* consumption and the progress of agricultural technology which introduces a degree of circularity into their construction. Thus the volume-based output figures must be subject to quite a wide margin of error and are not independent of output estimates based on population growth. In addition, the interpolation of gaps in the time series may have the effect of smoothing over fluctuations. They are shown in Table 3.5(c).

Land productivity

Dividing these estimates of output by estimates of land area gives a measure of land productivity. Table 3.6 shows agricultural land as estimated by contemporaries (with the exception of the figures for 1871 from the agricultural statistics), and these are used in constructing the land productivity estimates (Q/T) in Table 3.7(a) employing both the population and volume methods of calculating output. Contemporary estimates also form the basis for Clark's figures of land productivity for crops and livestock shown in Table 3.7(b), which suggest that the growth in livestock

Table 3.7 *Land productivity estimates, 1300–1860*

(a)	From population and volume based output estimates		
	Population method	Volume method	
	Index	£ per acre	Index
1700	100	1.91	100
1750	108	2.22	116
1800	115	2.64	138
1850	207	4.13	216

(b) Clark's estimates of output per unit area for southern England (wheat bushel equivalents)

	Crops	Livestock	All
c. 1300	3.05	1.04	4.1
c. 1850	6.73	6.56	13.2

(c) Cereal yields (bushels per acre)

	Lincolnshire (Wheat)	(WACY)	Norfolk & Suffolk (Wheat)	(WACY)	Herts (Wheat)	Hants (Wheat)	Wheat Index	WACY Index
c. 1300			14.9	11.5		10.8	79	115
c. 1550	9.5	8.0			9.0		57	80
c. 1600	11.7	9.9	12.0	8.5	12.2	11.0	72	92
c. 1650	15.8	10.0	14.5	9.3	16.0	12.9	91	96
c. 1700	15.6	10.7	16.0	9.2	17.0		100	100
c. 1750	20.0	13.5	20.0				123	135
c. 1800	21.0	15.8	22.4		24.0	21.0	136	158
1830s	22.9	20.0	23.3	21.0	21.6	21.6	138	205
1860	31.0		31.1		28.0	27.0	180	250

WACY = wheat, rye, barley and oat yields, weighted by crop proportions and crop price relative to wheat, for Lincolnshire, Norfolk and Suffolk only. Figures for *c.* 1600 and *c.* 1700 are distorted by the poor harvests of the 1590s and 1690s.

(d) Cereal yields in England 1801–71 (bushels per acre)

	1801	*c.* 1836	1871
Wheat	22	21	28
Barley	29	30	34
Oats	32	33	43
WACY	17	18	25

Counties used are: Bedfordshire, Buckinghamshire, Cambridgeshire, Cornwall, Derbyshire, Devon, Durham, Essex, Gloucestershire, Hampshire, Herefordshire, Huntingdonshire, Kent, Lincolnshire, Northumberland, Shropshire, Somerset, Staffordshire, Surrey, Sussex, Warwickshire, East Yorkshire, North Yorkshire and West Yorkshire.

productivity (QA/T) from the middle ages to the mid-nineteenth century was over sixfold compared with a doubling for crops (QC/T). It is impossible to produce a chronology of livestock productivity between these two benchmark dates although it is possible to construct a series of cereal yields per sown acre (QC/TC). A great variety of contemporary sources, including farm accounts, tithe accounts, and especially printed literature on farming, record yields per acre. Unfortunately there are not enough of these direct observations to construct a series of yields, and historians have once again been forced into teasing yield figures from sources that did not record them directly.

The first of these sources is probate inventories. Using a method developed by Overton and modified by Allen it is possible to estimate yields using valuations of standing and stored grain rather than physical quantities. The assumption is that those drawing up an inventory valued standing grain just before the harvest according to the price for which the grain would sell when it was harvested. This probably involved them in a calculation in which the value of the grain was computed as the post-harvest price per bushel multiplied by the yield in bushels per acre, minus the costs of harvesting the grain and the tenth which had to be given as tithe. Using this method yields have been calculated for Lincolnshire, Norfolk, Suffolk, Hampshire and Hertfordshire and the results are shown in Table 3.7(c). Cereal yields (QC/TC) are given as a 'weighted aggregate cereal yield index' for wheat, barley, rye and oats, which weights the crops according to their prices relative to wheat, and in proportion to the acreages sown. It is thus a single integrated measure of cereal yields.

Few inventories are extant after the mid-eighteenth century so the yield series is continued by using material from different sources. In the late 1760s and early 1770s the agricultural writer Arthur Young made several tours describing the farming activity in various parts of the country and his publications include reports of the yields he observed. However, there is some controversy over Young's capabilities as an agricultural reporter and the representativeness of his yield figures has been questioned. By the turn of the century much more reliable evidence is available, mostly from an enquiry initiated by the government into the harvest of 1801. The 1801 crop return records the acreages under five arable crops, but many of the parish returns also include information on yields. Thereafter, the tithe files include information on yields, and data are available from a series of private surveys mostly carried out by agricultural journalists. It is not until 1885 that yield information is available as part of the agricultural statistics. Table 3.7(d) shows cereal yields for a collection of English counties for 1801–71.

Yields per sown acre are recorded directly in medieval manorial accounts dating from the thirteenth century, and by combining the evidence from all

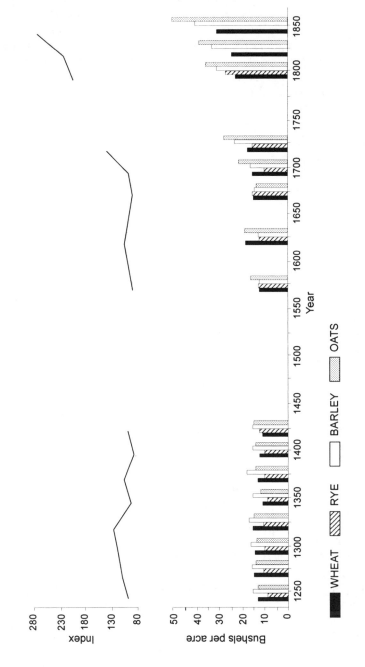

Figure 3.6 Cereal yields in Norfolk, 1250–1854. *Source:* Campbell and Overton (1993), 71.

these sources, albeit at the moment only for the counties of Norfolk and Hampshire, it is possible to chart the course in yields from 1250 to the present day. Figure 3.6 shows the yields of wheat, rye, barley and oats in the form of histograms for Norfolk, and also includes the index of weighted aggregate cereal yield described above. These estimates from inventories are calculated from the 'bottom up'; they are aggregations of many estimates for individual farms and therefore have the advantage that they can be related to other aspects of farm enterprises. Unfortunately inventories give no information on total arable acreages (they omit fallow) or on total farm acreages, so they cannot be used to calculate crop productivity measured in terms of the arable acreage (QC/TAR).

In comparison with crop productivity we have very little information on the productivity of the livestock sector, aside from Clark's benchmark estimates. Information on the weights of animals, on their yields of milk, meat, wool, leather and tallow, are rare before the late eighteenth century. Despite these difficulties, as we shall see, there are grounds for believing that increases in livestock productivity from the sixteenth to the nineteenth centuries were greater than those for crops. Some estimates consider that cattle weights rose by over 40 per cent in England over the eighteenth century, but a recent survey of the evidence concludes that there was little change in the sizes of cattle between the sixteenth century and the nineteenth century; the major change had already taken place between the thirteenth and the early sixteenth centuries; a finding partly corroborated by Scottish evidence. The cattle of the late seventeenth and early eighteenth centuries were comparable in size to modern animals. But weights and sizes for all animals could vary enormously: portraits of animals bred to almost fantastic sizes, which become common in the eighteenth century, are not necessarily exaggerated: they were just very untypical of the common beast.

Labour productivity

In comparison with land productivity, labour productivity is a relatively straightforward concept: it is simply the output per worker in agriculture, or Q/L where L is labour input. Such an index could be calculated from the 'bottom up' for particular farms using farm accounts, but no such exercise has yet been published. However, some national, 'top-down', estimates have been produced, using the indirect evidence of the proportion of the population engaged in agriculture. If a high proportion of the population are employed in agriculture it follows that labour productivity must be low, and, conversely, high labour productivity would be associated with a low proportion of the population working in agriculture. Thus if 75 per cent of the employed population is in agriculture, each worker produces enough food for themselves and for one third (25/75) of the requirements of one other person. When the proportion drops to 50 per cent then each agricultural worker produces enough food for himself or herself and for one other person. Wrigley has produced estimates of the proportion of the population working on the land, shown in Table 3.8(a), and also of the size of the 'rural agricultural population', shown in Table 3.8(b). These rural population estimates are related both to the population-based output estimates in Table 3.8(b) and to the volume-based output figures in Table 3.8(c).

Clark has derived estimates of labour productivity using a different method. He compares flat-rate wages for husbandry tasks (the daily wage for a task) with piece-work rates for the same task (payments for harvesting an acre of wheat or threshing a certain quantity of grain). Dividing one by the other gives an estimate of the labour input required for a particular agricultural task. Labour inputs for certain tasks remained constant, such as threshing, while others rose as yields rose, such as harvesting. He is thus able to estimate output per man day for given yields, so Table 3.8(d) shows output per man day associated with certain yield levels, and the dates at which these levels of yields are assumed to have prevailed. These figures are then used to estimate the proportion of the workforce engaged in agriculture at varying dates, and these figures, are shown alongside Wrigley's in Table 3.8(a).

Clark's figures imply that labour productivity was about 40 per cent higher than the Wrigley estimates for *c.*1600 (49 per cent versus 70 per cent of the population in agriculture), although both are agreed on the mid-nineteenth-century figure. The discrepancy may be due to the different methods of estimation, but the differences may also reflect a real change in the way labour was employed. Clark's figures are calculated on the basis of output per man day, whereas Wrigley's are based on proportion of the population in agriculture, which takes no account of the time workers actually spent working. Thus both sets of figures could be correct, and the difference between them could reflect the fact that the amount of work each person employed in agriculture performed was increasing.

Total factor productivity

Both land and labour productivity, however we may calculate them, are single factor productivities. That is they measure output in relation to only one input, land or labour. Each in isolation is only a partial guide to the efficiency of agricultural production as a whole. For example, it is possible to increase output per acre through increasing labour inputs, although the additional output will not be great enough to prevent a fall in labour productivity. Similarly, output per acre could go up in response to capital investment in land, by draining for example, although the additional output might not result in an increase in the return on capital. Total factor produc-

Table 3.8 *Labour productivity estimates, 1520–1871*

(a) Percentages of the population engaged in agriculture

Date	Wrigley	Clark
1520	76	
1600	70	49
1700	55	42
1800	36	33
1850	22	22

(b) 'Population' method

Date	Rural agricultural population (m.)	'Labour productivity'	Index (1700 = 100)
1520	1.82	1.32	71
1600	2.87	1.43	77
1670	3.01	1.65	89
1700	2.78	1.86	100
1750	2.64	2.34	126
1801	3.14	2.62	141
1831	3.38	3.45	185
1851	3.84	3.66	197
1871	3.35	4.81	259

(c) Labour productivity based on estimates of the volume of output (£ per head of the rural agricultural population)

	Productivity	Index (1700 = 100)
1700	14.5	100
1750	19.4	134
1800	24.5	170
1850	29.8	206

(d) Clark (southern England, output per man day versus yields)

Yield	Date	Total output per man day
10	1580, 1300	66
12		74
14		80
16		86
18		91
20		96
22	1790	100
24		104
26	1850	107
28	1860	110

Table 3.9 *Total factor productivity estimates, 1760–1870 (% change per annum)*

(a) Crafts ('output method')	
1761–1800	0.2
1801–1831	0.9
1831–1860	1.0

(b) McCloskey ('price method')	
1780–1860	0.45

(c) Huekel ('price method')	
1790–1815	0.2
1816–1846	0.3
1847–1870	0.5

(d) Mokyr ('price method')

(i) Using Williamson's wage data and Gayer-Rostow-Schwartz prices

	(1)	(2)
1797–1827	0.13	0.02
1797–1835	0.37	0.27
1805–1827	0.21	0.15
1805–1835	0.18	0.42

(ii) Using Bowley-Wood wage data and Gayer-Rostow-Schwartz prices

	(1)	(2)
1790–1820	0.32	−0.39
1820–1850	0.36	0.98

(1) Assuming the proportion of wages at 0.75 of factor prices and rents at 0.25.
(2) Assuming the proportion of wages at 0.33 of factor prices and rents at 0.67.

tivity attempts to measure output in relation to all inputs. The empirical difficulties of measuring single factors are multiplied when it comes to measuring all inputs and relating them to output. Moreover the calculation of total factor productivity also involves making a number of economic assumptions about the behaviour of the economy which may be unrealistic in the context of the early modern world. Crafts' estimates of total factor productivity are shown in Table 3.9(a). These are based on estimates of physical output and rely on evidence of output, price levels, and the relative shares of rent and wages in national income. They also depend on a wide range of economic assumptions about the nature of the economy, such as the existence of perfectly competitive product and factor markets, constant returns to scale, and disembodied technological change.

An alternative method of calculating total factor productivity using prices avoids the difficulties of measuring physical output. The rates of growth of input prices are compared with the rates of growth of output prices and any decline in output prices not attributable to a decline in input prices must reflect lower costs induced by productivity change. The other estimates in Table 3.9 employ this method. Although this technique is a fairly simple one it still depends on several assumptions about economic behaviour. Mokyr's results shown in Table 3.9(d) show how sensitive the indices are to the choice of price and wage series and to the weights given to rents and wages in the calculations. While in theory total factor productivity is the best way of measuring the efficiency of agricultural production, and therefore the development of technical change, in practice it is fraught with difficulties which make the results of the exercise of dubious value. The important point is that a single measure of productivity, be it for land or labour, is in itself incomplete and may not be an accurate reflection of agricultural efficiency.

Summary of output and productivity trends

The results of these calculations of output and productivity are shown as annual growth rates in Table 3.10, simplified as index numbers in Table 3.11, and graphed in Figure 3.7. Before the eighteenth century the only available output figures are those based on population, so naturally the rate of growth in agricultural output reflects the rate of population growth. This is probably exaggerated, since the evidence of prices and wages suggests that output per head was not constant, but the extent of the exaggeration is impossible to estimate. Likewise, if consumption per head rose after 1650 (which again, given the evidence of prices and wages, it probably did), output may not have fallen as the growth rate for 1650–1700 indicates. For the periods after 1700 the various estimates are agreed on one point: the fastest rate of growth was in the first few decades of the nineteenth century, rather than at any period in the eighteenth century. For the eighteenth century both the population-based estimates of output and those derived from the volume of output agree that growth was more rapid during the second half of the century (0.55 and 0.81 per cent per annum) than it was during the first half (0.38 and 0.48 per cent per annum). But the demand equation technique of Crafts reverses this finding: rising prices and falling real incomes after 1760 would have reduced consumption per head, and so, according to this method, growth was more rapid in the first half of the century. However, for the course of the eighteenth century as a whole the

Table 3.10 *Population, output and productivity, 1520–1850 (% change p.a.)*

	Population		Land area				Output			Land productivity			Labour Productivity	
	Total	Non-agricultural	Total	Arable	Sown arable	Meadow & pasture		(1)	(2)	(3)	(4)	(5)	(6)	(7)(8)
1520–1600	0.77	0.95	–	–	–	–	0.77	–	–	–	–	–	0.10	–
1550–1600	0.62	–	–	–	–	–	0.62	–	–	–	–	0.46	–	–
1600–1650	0.48	–	–	–	–	–	0.48	–	–	–	–	0.47	–	–
1600–1670	0.27	0.66	–	–	–	–	0.29	–	–	–	–	–	0.20	–
1650–1700	-0.07	–	–	–	–	–	-0.03	–	–	–	–	0.18	–	–
1670–1700	0.05	0.49	–	–	–	–	0.07	–	–	–	–	–	0.40	–
1700–1750	0.26	0.64	–	–	–	–	0.38	0.48	–	0.15	0.30	0.42	0.46	0.59
1700–1760	0.33	–	–	–	–	–	0.36	–	0.60	–	–	–	–	–
1750–1800	0.82	1.14	–	–	–	–	0.55	0.81	–	0.14	0.38	0.20	0.23	0.47
1760–1800	0.86	–	–	–	–	–	0.62	–	0.44	–	–	–	–	–
1800–1830	1.45	1.97	–	–	–	–	1.18	–	1.18	–	–	0.04	0.92	–
1830–1850	1.16	1.33	–	–	–	–	0.94	–	–	–	–	0.90	0.30	–
1700–1800	0.55	0.89	0.32	0.25	0.30	0.38	0.46	0.65	0.53	0.15	0.34	0.31	0.35	0.54
1750–1850	1.07	1.42	–	–	–	–	0.82	0.81	–	0.66	0.65	0.35	0.45	0.43
1800–1850	1.36	1.71	-0.09	0.57	0.78	-0.70	1.08	0.81	–	1.17	0.92	0.47	0.67	0.39

Notes

(1) Population based method
(2) Volume based method
(3) Crafts' estimates
(4) Derived from population based output method
(5) Derived from volume based output method
(6) Wheat yields from counties with inventory data
(7) Derived from population based output method
(8) Derived from volume based output method

Table 3.11 *English agricultural output and productivity, 1300–1850*

	1300	1600	1700	1750	1800	1850
A. Output						
Output (population method)		80	100	121	159	272
Output (volume method)			100	127	191	285
Output (demand equation method)			100	143	172	244[a]
B. Area						
Arable area			100		128	170
Sown arable area			100		135	199
Meadow and pasture			100		147	103
Total area			100		138	132
C. Land productivity						
Land productivity (population)			100		115	207
Land productivity (volume)			100		138	216
Crop productivity[b]	3.05					6.73
Livestock productivity[b]	1.04					6.56
Wheat yields[c]	79	72	100	123	136	180
Cereal yields[d]	115	92	100	135	158	250
D. Labour productivity						
Labour productivity (population)		77	100	126	141	197
Labour productivity (volume)			100	134	170	206

[a] 1831
[b] Clark's estimates in wheat bushel equivalents
[c] Hampshire, Hertfordshire, Lincolnshire, Norfolk, Suffolk (the 1300 average for Norfolk and Hampshire only)
[d] Norfolk and Suffolk

three estimates are remarkably similar at 0.46, 0.53 and 0.65 per cent per annum. This is also reflected in Table 3.11 where all three output estimates show remarkable similarity over the entire period 1700–1850. By the nineteenth century, growth had accelerated, and, as Figure 3.7(a) shows, was outstripping growth in the sown arable acreage.

The nineteenth century also witnesses the most rapid growth in land productivity: the increase over the eighteenth century is particularly pronounced for the population- and volume-based measures of land productivity (Q/T), at 1.17 and 0.92 per cent per annum for the period 1800–50, shown in Figure 3.7(b) and Table 3.11. Wheat yields grew less rapidly (0.47 per cent per annum), suggesting that livestock productivity

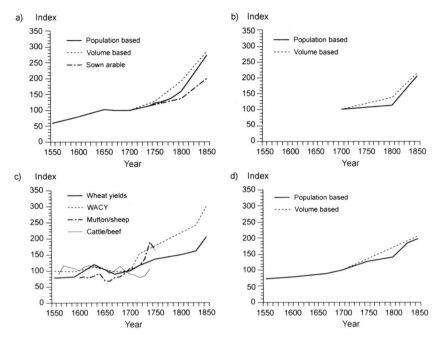

Figure 3.7   (a) Agricultural output and sown arable in England, 1550–1850, (b) land productivity in England, 1700–1850, (c) grain yields and estimates of livestock output in England, 1500–1850, and (d) labour productivity in England, 1550–1850. *Sources:* See Tables 3.5–3.8, 3.10 and 3.11.

may have been growing more quickly than cereal yields, although the yields of other grains were growing more rapidly than those of wheat. Furthermore, although cereal yields were rising from 1700, the arable area was also increasing, and during the first fifty years of the nineteenth century the sown arable area was increasing more rapidly than cereal yields (at 0.78 per cent per annum). The growth in wheat yields shows some fluctuation from the mid-sixteenth century; growth rates were high (over 0.46 per cent per annum) during the period of population growth, but fell back as population growth ceased from the mid-seventeenth century.

Labour productivity appears from the figures in Table 3.10 to have grown only slightly in the sixteenth century (only 8 per cent in total from 1520 to 1600). In fact it is likely that labour productivity fell during this period because the figures are based on the proportion of the population engaged in farming. If the average number of hours per annum each worker in agriculture spent working increased over the period, then the denominator (L) in the productivity equation (Q/L) increases and labour productivity falls. The downward trend in real wages suggests that this was the case, as does the general evidence of increased employment suggested by, among other factors, a reduction in the number of holidays. It seems likely that the rise in labour productivity began in the mid-seventeenth century, and although the two estimates in Figure 3.7(d) differ, thereafter the trajectory was upward.

Increasing output

It is now necessary to consider the ways in which this increase in agricultural output might have been achieved. In answering this question a distinction that at first sight seems helpful is between improvements in output brought about through the extension of the cultivated area, and improvements in output brought about by a rise in land productivity through technological change. Extensions to the cultivated area are a less significant cause of increasing output than productivity change because the limits of the land available would soon be reached, whereas technological change offers far more possibilities for 'revolutionary' increases in output. Relatively little new land was available in the sixteenth century, when at least three quarters of the land farmed today was being cultivated, yet between *c.* 1500 and the present day average wheat yields have risen over twelvefold. In practice however, the distinction between output increases caused by rises in yields and by extensions of the cultivated area is difficult to draw. In certain situations extending the cultivated area was only possible in conjunction with technological change, by using new techniques of cultivation, or by growing new crops. In any case the 'cultivated area' is not such a simple concept as it might first appear. Even in the sixteenth century very little land was of no agricultural use whatsoever, except barren mountain with no vegetational cover. In a sense therefore, expanding the cultivated area really means increasing the intensity of land use by changing the composition of agricultural output. Thus the discussion of increases in agricultural output will be first in terms of changes in the composition of output and secondly in terms of sustained increases in output per acre. Before this, however, it is necessary to look at another way in which food supply could have increased; through an increase in net imports.

Table 3.5(a) shows some estimates of net imports as a percentage of the total amount of agricultural produce consumed in the country at various dates, and Figure 3.8 shows the more reliable figures of net exports of grain. Trade figures only become reliable after 1660 when the grain trade in particular was well documented. Tracing the movement of other commodities is more difficult, as is the movement of agricultural produce between Ireland, Scotland and England. By the early decades of the nineteenth century some 70 per cent of English food imports were from Ireland, where living standards declined as those in England rose. Although the English

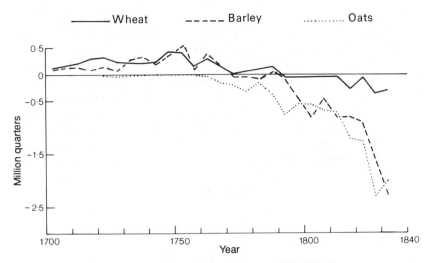

Figure 3.8 Net exports of wheat, barley and oats, 1700–1830 (5-year averages).
Source: Mitchell and Deane (1962), 96–7.

export achievement in the eighteenth century was in one sense modest, with a maximum of only 8 per cent of produce being exported, it did provide a buffer, albeit a small one, when demand subsequently increased with population growth. From the turn of the nineteenth century imports increasingly became important in meeting the demand for food, so that by mid-century they were accounting for over one sixth of consumption. However, the 'agricultural revolution' was not based on food imports.

Changes in the composition of output

Changes in the composition of output can be considered under six headings: land reclamation or the improvement of land quality through capital investment, changes in the ratio of grassland to arable, the reduction of fallows, the introduction of fodder crops, changes in the balance between different food crops, and regional specialisation. Evidence for the first of these, the improvement of land quality, becomes commonplace from the mid-sixteenth century. Historical geographers have charted the draining of marsh and fen, the continued clearing of woodland, the reclamation of the heathlands, and the reclamation of upland 'waste'. The most dramatic of these developments was the attack on the marshes and the fens, especially from the seventeenth century. The most spectacular reclamation was in the peatlands of the southern fenland in eastern England, but drainage also took place in the Somerset Levels, in Hatfield Chase, the Isle of Axholme, on the coastlands of the Thames estuary in Essex and Kent, and the meres of Holderness in east Yorkshire (Figure 2.10). The fenland, which in the early seventeenth century had supported an economy based on fishing and fowling, had, by the second half of the eighteenth century, become some of the most fertile arable land in the country. However, continual shrinkage of the newly drained peat meant that the process of draining became a protracted one, since, as the land sank, water had to be pumped up to the channels draining it away. By the eighteenth century the pumping was done by windmills, and these were then replaced by steam pumps from the 1820s. Although major drainage activity was carried out in the mid-seventeenth century, the peak of drainage activity was in the late eighteenth and early nineteenth centuries. This new land was used to grow conventional arable crops, but in some areas it became the focus for the intensive cultivation of industrial crops such as rape, flax and hemp. As Sir John Maynard put it in 1650, 'Rape, Cole-seed, and Hemp, is but a Dutch Commodity, and but trash and trumpery.' No quantitative estimates are available of the extent of the increase in the area of crops and grass brought into production by this activity but it must have been considerable, amounting to over 6 per cent of the total land area of England and Wales, perhaps extending the arable acreage by some 10 per cent.

It is difficult to measure the loss of woodland from the sixteenth to the nineteenth centuries, although recent research has tended to stress the extent to which woodlands were preserved rather than destroyed during these centuries. In 1350 roughly 10 per cent of England was wooded; by the middle of the nineteenth century the figure was around 5 per cent, although both estimates are very approximate. Locally, woodland losses could be severe, as in Norfolk which lost three-quarters of its medieval woods between 1600 and 1790, but most of the great woodland areas, such as the weald of Kent and Sussex, remained intact. Historians have argued that woodland losses were severe because of the demands of industry; their error is in forgetting that trees and wood regenerate. Much wood for fuel was provided by coppicing, and although demand for standard trees for the construction of both buildings and ships was heavy there were extensive replanting schemes from the late seventeenth century onwards. The complete removal of woodland was an expensive business, as an account of woodland clearance in the mid-nineteenth century illustrates. The clearing of 3,000 acres of Wychwood forest near Woodstock in Oxfordshire firstly involved driving ten miles of new roads at a cost of £700 a mile. Next the deer of the forest were killed. It cost over £5,000 to clear brushwood, fell trees, and prepare wood and timber for sale (although timber sales were to raise over £16,000) and more than £6,000 was spent in grubbing up the roots of the trees once they had been cleared. Finally new fields were fenced and seven farms laid out with new farmhouses. All in all the net outlay on the reclamation project was over £10,000.

The other areas that were reclaimed were rough pastures and heathlands. Pastures were usually in upland areas and required stone-clearing and wall-building, deep ploughing, draining and liming to improve them. Some moorlands, especially in the south-west, but also on the Yorkshire Wolds, were improved by paring and burning (also known as 'denshiring'), when the turf was removed by a breast plough, piled into heaps, and burnt, the ashes being incorporated in the newly ploughed land. Upland wastes were gradually encroached upon from the sixteenth century onwards, but when pressure on land eased, as in the early eighteenth century for example, land reverted back to waste. The real attack on upland wastes came in the century after 1750, and particularly in the first two decades of the nineteenth century.

While undrained marsh produced little beyond fish and fowl, heathlands probably yielded rough grazing and thus supported animals, although the carrying capacity of such land would have been low. When they were reclaimed the transformation could be spectacular, and many former heathland areas were the locus of intensive 'high farming' systems by the 1840s. Reclamation was associated with the introduction of new crops and crop rotations (discussed in more detail below). Root crops, particularly turnips, coupled with the extensive use of marl and lime were responsible for turning heathlands and some downlands (see Figure 2.6) into productive land growing wheat and barley, with fodder crops supporting large numbers of animals. They did this by taking nutrients from the soil (up to five times the amount of cereal crops) and, since their roots were deeper in the ground, from a different level in the soil. These nutrients could then be recycled, either as manure, or through crop residues left in the soil. Aside from turnips, other new crops were involved in this process of heathland reclamation, including sainfoin, an 'artificial grass'. (Incidentally, these 'artificial grasses' were neither artificial nor, in many cases, grasses; they were so called because they were sown from seed and were sometimes not indigenous.) The reclamation of heathlands in this way is not however quite the same as the reclamation of woodland. Soils under woodland could be inherently fertile, but those under heath were not, and as a French commentator put it in 1784:

The fertility of this land is entirely artificial: the factitious vegetable bed is perhaps no more than eight or ten inches deep, and a few years of bad management would make it as impoverished today as it was before. The rotation of crops is: 1. wheat; 2. turnips; 3. barley and clover; 4. clover that is cut for the first year, left for the second and sometimes the third, to be grazed by the flocks.

The reclamation of heathlands in this way is a classic case of land reclamation associated with the innovation of new crops. An anonymous author writing about this improved agriculture in Norfolk in 1752 commented, 'We sow on these improved farms five times as many acres of wheat, twice as many of barley.' Evidence is also available from probate inventories from Norfolk and Suffolk to show that between 1660 and 1730 the average barley acreage per farm increased 3.8 times. Some of this increase might have been due to increasing farm size, but even so it suggests that the cereal acreage was expanding.

Accurate figures of the proportion of the country's land which was reclaimed, in the sense that it was converted to more productive systems of land use, are impossible to estimate. In the early seventeenth century only some 6 per cent of Leicestershire was classed as 'waste', but in Devon 20 per cent of the land in 1600 was still under natural vegetation. Although some reclamation took place during the first half of the seventeenth century, by the end of the century Gregory King reckoned that 'heaths, moors, mountains and barren land' still comprised about a quarter of England and Wales. It is certain that the pace of reclamation increased from the mid-eighteenth century onwards. The estimates in Table 3.6 suggest that over the course of the eighteenth century the area of arable, meadow and pasture grew by 38 per cent. Reclamation reached a peak during the Napoleonic Wars when agricultural prices were rising at an unprecedented rate. This is reflected in estimates of capital investment in agriculture. Feinstein considers that investment in agriculture as a percentage of gross rentals rose from around 6–7 per cent in the 1760s and 70s to 11 per cent in the 1790s, and 16 per cent during the period 1801–10. The Board of Agriculture considered there were over six million acres of waste in England around 1800, and most of that was upland waste. As the President of the Board of Agriculture, Sir John Sinclair, put it in 1803, 'Let us not be satisfied with the liberation of Egypt, or the subjugation of Malta, but let us subdue Finchley Common; let us conquer Hounslow Heath, let us compel Epping Forest to submit to the yoke of improvement.' The area of wasteland in England enclosed by act of parliament between *c.* 1750 and 1850 shown in Figure 4.3(b) was about 7 per cent of the country's area. By 1873 only some 6–7 per cent of the country remained as waste.

Changes in the ratio of arable to grassland can increase output since tillage crops are consumed directly by human beings whereas fodder crops are eaten by animals before they can be made available as human food. Since energy in the fodder crops is lost in keeping the animal alive before it can be consumed by human beings it follows that animals provide less human food per acre than do tillage crops, although the difference in the land productivity of different usages varies, depending on the nature of the livestock enterprise. Modern studies show that the cereals produce six times as many calories per acre as does milk, the most efficient animal product. Thus, overall land productivity (Q/T, where output is measured in calories) could be increased considerably simply by changing land use from fodder crops to grain.

There were of course constraints to this strategy: not all grassland could be converted because animals were needed for draught and to recycle nitrogen through their manure. There were also many areas where topography, soils and climate were such that arable crops could not replace grass: in wet lowland meadows or in upland areas with poor soils and steep slopes for example. Much clayland was under permanent pasture because it was too wet to cultivate for cereal crops. Drainage was achieved by ploughing the land in ridge and furrow, but in many years the furrows would be waterlogged. Widespread underdrainage did not occur until the second half of the nineteenth century, but there were a number of strategies for improving the quality of the soil, such as deep ploughing, paring and burning, and the addition of lime.

Unfortunately sources do not enable the balance between arable and pasture to be charted with any certainty. It seems that in the sixteenth century there was a general switch towards arable at the expense of pasture under the stimulus of rising population and differential price movements for livestock and crops. The century after 1650 saw the opposite trend: the extension of pasture at the expense of arable. This is very evident during the early period of parliamentary enclosure in the Midlands for example, but it was also a phenomenon of non-parliamentary enclosure. Table 3.6 suggests that the acreage of meadow and pasture grew by some 47 per cent over the eighteenth century compared with 38 per cent for arable, although the figures also show the opposite trend from 1800 to 1830, with a 30 per cent decline in meadow and pasture and a rise in the sown arable acreage of some 47 per cent.

No data are available to give an accurate figure of the balance between arable and grassland before 1866 when the modern agricultural statistics were inaugurated. Before then it is, however, possible to provide some material on the proportions under various crops by county, which help in examining the remaining ways in which the composition of output changed: through the reduction in fallows, the introduction of fodder crops, and changes in the balance of food crops. Table 3.12 pulls together data from county studies of probate inventories, where it is possible to provide comparable data on the proportions of arable under cereal crops and pulses. These proportions are carried forward to the nineteenth century using data from the 1801 crop return, the tithe files and the agricultural statistics for 1871. Where they are available the percentages under turnips and seeds are also given, although they are shown in italics because the percentages are calculated from a different total. Table 3.13 shows the proportion of arable under various crops from the 1801 crop return, and Table 3.14 for the tithe files and the 1871 statistics. The data in Tables 3.13 and 3.14 are not directly comparable because the percentages are calculated from different groups of crops, so Table 3.15 shows the percentages under five crops which are recorded in all three sources.

Table 3.12 *Crop percentages for selected areas, 1530–1871*

	Wheat	Rye	Barley	Oats	Pulses	Turnips	Seeds
Cornwall 1600–1620	49	1	13	38	–		
East	48	2	4	46	–		
Centre north	45	0	16	39	–		
Centre south	51	1	13	35	–		
West	50	2	18	30	–		
Cornwall 1680–1700	36	0	42	22	–		
East	32	0	38	31	–		
Centre north	24	0	48	29	–		
Centre south	37	0	51	12	–		
West	52	1	32	15	–		
Cornwall 1801	39	0	38	24	0	6[a]	
Cornwall 1871	35	0	36	29	0	9[b]	39[b]
Hertfordshire							
1540–1579	28	15	18	21	18		
1580–1609	31	12	14	26	17		
1610–1639	29	7	17	24	23		
1640–1669	32	5	21	21	21		
1670–1699	26	3	23	20	23		
c.1836	42		36	13	9	13[b]	20[b]
1871	42	0	30	16	11	12[b]	17[b]
Kent 1600–1620	42	1	23	21	14		
Weald	38	3	2	50	7		
Sandstone	39	0	31	13	17		
Downland	43	1	25	17	15		
North Kent	46	0	33	4	17		
Kent 1680–1700	31	0	22	31	16		
Weald	30	0	8	60	2		
Sandstone	32	0	26	32	10		
Downland	23	0	34	19	23		
North Kent	39	1	21	12	28		
Kent 1801	41	0	14	22	24	7[a]	
Kent *c*. 1836	44		21	18	18	12[b]	19[b]
Kent 1871	44	0	16	21	19	9[b]	17[b]

Table 3.12 (*contd.*)

	Wheat	Rye	Barley	Oats	Pulses	Turnips	Seeds
Lincolnshire 1530–1600	19	5	46	4	27		
Fenland	10	2	57	0	30		
Marshland	32	5	31	2	31		
Wolds and heath	10	4	59	9	16		
Clay and misc.	22	7	38	3	31		
Lincolnshire 1630–1700	24		36	9	31		
Fenland	33		16	7	43		
Marshland	24		42	1	32		
Wolds and heath	18		43	23	16		
Clay and misc.	21		41	3	34		
Lincolnshire 1801	30	2	27	29	13	*19[a]*	
Fenland	30		14	43	13		
Marshland	41		14	14	32		
Wolds and heath	25		41	26	8		
Clay and misc.	32	4	27	23	14		
Lincolnshire *c.*1836	46		19	25	10	*12[b]*	*24[b]*
Lincolnshire 1871	49	0	25	17	9	*15[b]*	*17[b]*
Norfolk and Suffolk							
1584–1599	21	19	43	7	10		
1628–1640	24	10	41	12	13		
1660–1699	25	9	40	8	18	*1[b]*	
1700–1739	23	6	47	10	14	*8[b]*	*3[b]*
*c.*1836	47		43	6	4	*20[b]*	*24[b]*
1871	43	2	38	6	13	*15[b]*	*19[b]*
Oxfordshire uplands							
1590–1640	14	4	61	7	15		
1660–1730	27	0	49	4	20		
East Worcestershire							
1540–1599	22	17	27	10	21		
1600–1660	18	16	24	15	24		
1670–1699	28	5	29	6	28		
1700–1750	35	2	26	8	28		
1801	45	2	23	9	21	*7[a]*	

Percentages for turnips and clover are italicised because they are calculated on a different basis from the other figures in the row.

[a] Turnip acreage as a percentage of the acreage of cereals, pulses and turnips
[b] Turnip and seeds (including clover) acreages as a percentage of the acreage of cereals, pulses, turnips and seeds

Table 3.13 *Crop proportions in the 1801 crop return (percentages)*

	Wheat	Barley	Oats	Rye	Potatoes	Turnips	Peas/ Beans
Bedfordshire	32	21	17	2	1	7	20
Buckinghamshire	38	17	17	0	0	4	23
Cambridgeshire	26	23	34	2	1	4	11
Cheshire	36	5	51	0	7	0	1
Cornwall	35	34	21	0	5	6	0
Cumberland	10	18	55	1	6	8	3
Derbyshire	31	13	42	0	2	6	5
Durham	37	5	42	1	3	8	4
Essex	41	17	20	1	2	6	15
Gloucestershire	36	25	15	0	2	10	12
Hampshire	36	27	20	0	1	9	5
Herefordshire	45	20	10	1	2	6	16
Kent	37	13	20	0	1	7	22
Lancashire	20	8	60	0	8	1	3
Leicestershire	28	26	25	0	2	11	9
Lincolnshire	24	21	23	1	2	19	10
Middlesex	36	16	11	2	5	4	26
Monmouthshire	42	23	20	0	3	4	8
Northamptonshire	28	27	17	1	1	9	18
Northumberland	25	14	43	1	2	11	4
Rutland	21	32	18	1	1	13	14
Shropshire	39	23	22	1	2	7	8
Somerset	44	18	14	0	5	7	11
Staffordshire	32	19	35	1	2	7	5
Surrey	32	18	26	1	2	11	11
Sussex	41	12	32	0	1	7	7
Warwickshire	38	23	19	0	1	8	12
Wiltshire	39	29	16	0	2	6	8
Worcestershire	42	19	10	1	1	7	19
Yorkshire East	28	14	31	2	1	14	11
Yorkshire North	27	9	38	3	3	15	6
Yorkshire West	33	13	35	1	3	8	8
All	32	19	26	1	2	8	11

Only counties with more than 10 per cent of their area covered by the returns are included

Table 3.14 *Crop proportions* c. *1836 and in 1871 (percentages)*

	Wheat		Barley		Oats		Pulses		Turnips		Seeds		Fallow	
	1830s	1871	1830s	1871	1830s	1871	1830s	1871	1830s	1871	1830s	1871	1830s	1871
Bedfordshire	23	35	13	19	7	6	14	18	11	8	17	10	16	5
Berkshire	24	27	16	17	8	12	6	10	18	16	23	16	5	2
Buckinghamshire	23	30	15	16	10	12	7	13	14	11	21	15	7	3
Cambridgeshire	23	40	11	17	14	11	6	11	5	5	28	13	11	4
Cheshire	31	21	4	3	32	27	0	3	3	5	9	40	16	1
Cornwall	–	16	–	16	–	13	–	0	–	9	–	39	–	7
Cumberland	–	10	–	4	–	28	–	0	–	13	–	42	–	2
Derbyshire	24	23	6	11	24	20	2	3	7	8	21	30	14	5
Devon	–	21	–	15	–	15	–	1	–	15	–	27	–	6
Dorset	–	21	–	19	–	10	–	4	–	20	–	23	–	3
Durham	22	21	5	8	22	19	3	4	5	11	22	26	21	12
Essex	26	35	19	19	6	7	9	15	7	5	18	13	16	7
Gloucestershire	–	29	–	13	–	5	–	9	–	13	–	28	–	2
Hampshire	21	23	15	14	6	13	2	4	16	19	33	23	4	4
Herefordshire	–	33	–	12	–	7	–	9	–	14	–	20	–	4
Hertfordshire	23	28	20	20	7	11	5	8	13	12	20	17	12	5
Huntingdonshire	24	37	17	17	6	8	11	16	3	3	16	11	21	7
Kent	28	32	13	12	11	15	11	13	12	9	19	17	6	2
Lancashire	–	18	–	5	–	26	–	3	–	5	–	41	–	2
Leicestershire	–	29	–	20	–	13	–	10	–	9	–	16	–	4
Lincolnshire	25	33	10	16	14	11	6	6	12	15	24	17	8	2
Middlesex	–	29	–	7	–	16	–	11	–	8	–	27	–	2
Monmouthshire	–	24	–	14	–	9	–	3	–	12	–	32	–	6
Norfolk	24	27	22	24	3	4	2	3	24	19	25	22	2	1
Northamptonshire	–	31	–	21	–	7	–	13	–	11	–	13	–	4
Northumberland	19	13	5	11	23	20	2	3	6	17	26	29	17	6
Nottinghamshire	–	28	–	18	–	8	–	9	–	14	–	19	–	5
Oxfordshire	24	25	17	21	7	9	7	10	14	16	24	17	7	2
Rutland	20	25	19	27	6	8	11	7	13	17	23	14	8	3
Shropshire	24	27	10	17	15	8	1	4	10	16	24	25	14	3
Somerset	–	29	–	14	–	8	–	7	–	14	–	25	–	3
Staffordshire	22	26	13	15	13	14	1	4	12	13	28	24	10	4
Suffolk	24	29	23	25	2	3	6	13	14	11	19	15	10	3
Surrey	23	27	9	12	14	15	5	7	9	12	25	20	13	7
Sussex	23	30	6	7	19	19	2	7	7	10	25	20	14	6
Warwickshire	22	34	12	13	11	6	7	16	8	9	27	19	13	4
Westmorland	–	4	–	6	–	30	–	0	–	15	–	42	–	2
Wiltshire	–	26	–	18	–	9	–	6	–	17	–	21	–	4
Worcestershire	–	37	–	11	–	4	–	17	–	8	–	18	–	4
Yorks. East	22	25	10	13	17	16	6	6	10	17	21	19	14	4
Yorks. North	24	19	8	18	17	17	5	4	9	15	19	20	17	7
Yorks. West	23	24	14	17	12	13	3	6	13	13	22	23	12	4
All	23	26	13	15	12	13	5	8	11	12	22	23	12	4

Table 3.15 Crop proportions in 1801, c. 1836 and in 1871 (percentages)

	Wheat			Barley			Oats			Peas/Beans			Turnips		
	1801	1836	1871	1801	1836	1871	1801	1836	1871	1801	1836	1871	1801	1836	1871
Bedfordshire	33	33	41	21	19	22	18	11	7	21	20	21	7	17	9
Buckinghamshire	38	34	37	17	21	19	17	15	15	23	10	16	4	20	13
Cambridgeshire	27	40	47	23	19	20	35	23	13	11	10	13	4	9	6
Cheshire	39	40	36	5	5	5	55	46	47	1	0	5	0	4	8
Derbyshire	32	38	35	13	10	17	43	38	31	5	4	5	6	11	13
Durham	39	39	34	5	8	13	44	39	30	4	4	6	8	9	17
Essex	32	39	43	17	28	23	20	9	9	16	13	18	6	11	6
Hampshire	37	34	31	28	25	19	21	10	18	5	4	6	10	27	26
Kent	38	37	39	13	18	14	20	14	18	22	15	17	7	16	11
Lincolnshire	24	37	40	22	15	20	23	21	14	11	8	7	20	18	18
Northumberland	26	35	20	15	9	18	44	41	31	5	3	5	11	12	26
Rutland	21	29	37	33	28	32	18	9	10	15	16	8	13	19	20
Shropshire	40	40	30	23	17	24	22	25	11	8	2	6	7	17	22
Staffordshire	32	36	29	20	21	21	36	21	20	5	2	6	7	20	17
Surrey	33	38	37	18	15	16	27	23	21	11	8	10	11	15	17
Sussex	41	40	41	12	11	10	32	33	26	7	3	10	8	13	14
Warwickshire	38	38	44	23	20	16	20	17	8	12	12	21	8	13	11
Yorkshire East	28	34	33	14	16	17	32	26	20	11	9	8	14	16	21
Yorkshire North	29	38	26	9	12	24	40	27	23	7	8	6	15	15	20
Yorkshire West	34	36	33	13	21	24	36	19	18	8	5	8	9	20	18
Average	34	37	36	17	17	19	30	23	20	10	8	10	9	15	16

Reductions in fallows increased output and both overall land productivity (Q/T) and the productivity of the arable ((QAF + QC)/TAR) because unproductive fallow was replaced by a crop. As we have seen in the previous chapter, fallows were essential in controlling weeds and allowing nitrification to take place; in a conventional three-field system one third of the arable would have been under fallow. Their replacement by a growing crop was possible because certain crops, such as turnips, controlled weeds while growing, and in the process conserved supplies of nitrogen. Turnips grew quickly and could smother weeds with their large leaves. If they were grown in rows, and hoed, then weeds could be controlled, and hoeing made moisture available to the crop. Although a bare fallow sees the addition of some nitrogen from the atmosphere by bacterial action, it also sees the loss of nitrates (which are water soluble) by leaching. The replacement of the bare fallow by a root crop would reduce this leaching and intercept the nitrogen that otherwise would be lost. Plants with large leaves mean more water is lost through transpiration than through drainage and so more nitrogen is retained. Furthermore if the roots were fed to livestock *in situ* then soil nitrogen would be recycled efficiently. The first truly reliable estimates of the fallow acreage are not available until the tithe files of the 1830s. Gregory King estimated the proportion of fallow in England and Wales in the 1690s at 20 per cent of the arable (Table 3.6). The fallow proportion had declined by 1800, but the fall was more rapid subsequently; at 12 per cent in the 1830s, and 4 per cent by 1871 (Tables 3.6 and 3.14).

Root crops were important both in reclaiming light land and in replacing fallows but they were also important (along with other new fodder crops such as clover) because they were a higher-yielding form of fodder than the grazing from permanent pasture. Thus the introduction of fodder crops permitted the arable area to expand at the expense of permanent pasture, since the relatively low-yielding pasture was replaced by the relatively higher-yielding fodder crops. The exact difference in yield (in terms of food-value) is hard to estimate but in the early years of the present century an average turnip crop gave 70 per cent more starch per acre than an average hay crop and 40 per cent more protein; clover hay 20 per cent more starch per acre and 80 per cent more protein.

Root crops, particularly the turnip, were recorded as a garden crop in sixteenth-century England: indeed it is possible that the Romans introduced them to Britain. They appear as market garden crops for human consumption in the late sixteenth century and were being grown as animal fodder on a handful of farms by the 1620s and 30s. A study of the innovation of turnips in Norfolk and Suffolk (Figure 3.9) has shown that under 1 per cent of farmers were growing the crop from the 1630s through to the 1660s. Thereafter the proportion of farmers growing turnips rises, to 20 per cent in the 1680s, 40 per cent in the 1700s and 50 per cent in the 1720s. Despite

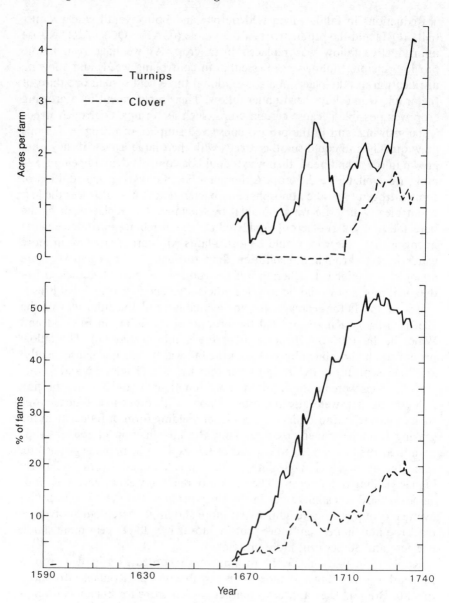

Figure 3.9   The diffusion of turnips and clover in Norfolk and Suffolk, 1584–1735. *Source:* Overton (1985), 208.

their popularity, turnips were not an essential part of crop rotations until the middle decades of the eighteenth century and in many cases do not seem to have been grown for their 'roots' at all but for their green tops. Even in the main centre of their cultivation, Norfolk and Suffolk, they only constituted 8 per cent of the arable acreage (excluding fallow) during the period 1700–39 (Tables 3.12 and 3.20). It also appears that turnips were sown after the harvest, in August, rather than in the spring. They were neither drilled nor hoed, and would not have acted as a cleaning crop as they were to later.

Thus despite the appearance of turnips in the seventeenth century we should not be misled into thinking that the crop was having a major impact on output and productivity at that time. The main period of the diffusion of turnip husbandry, which involved the spring sowing and hoeing of turnips, both in Norfolk and Suffolk and in the rest of the country, was after 1750. The earliest available national statistics are those from the 1801 crop return shown in Table 3.13 (unfortunately there are insufficient data in 1801 for Norfolk and Suffolk). The tithe files of the 1830s also show the proportion of arable under turnips (Table 3.14) but comparisons with 1801 should be made using Table 3.15, in which the percentages are directly comparable. These tables give an indication of both the geography and chronology of the introduction of turnips. The crop still had much ground to cover in 1801, contributing to 9 per cent of the cropped acreage (excluding rotational grasses), but it reached some 15 per cent by the 1830s. Although some counties showed little change between the two surveys, in others the spread of the crop was very rapid; from 4 to 20 per cent in Buckinghamshire for example. Table 3.14 also suggests an inverse relationship between the proportions of fallow and turnips. In fact the correlation on a county basis between the proportions of land under fallow and under turnips is a remarkable –0.84: in three counties using data from individual tithe districts the coefficient reaches –0.9. This is the clearest evidence we have that turnips replaced fallows.

Towards the end of the eighteenth century new root crops began to be grown to supplement and replace turnips. Turnips were subject to 'finger and toe' disease (the modern gardener recognises this as club root) and to attacks from the turnip fly. They were also susceptible to frosts; in the 1780s for example a Suffolk farmer reported that his turnip crop was destroyed by frost once every six or seven years. The most important introduction was the Swedish turnip (soon abbreviated to swede), which was brought to England by a Kentish farmer in 1767, and spread quickly in the early years of the nineteenth century. Another introduction was the mangel, which spread rather later than the swede and was cultivated on heavier soils. As well as being more hardy both these new crops provided more nutritious fodder than did the common turnip.

The introduction of turnips and other new fodder crops is one example of a change in the crop mix which could have an effect on output. Changes in the mix of crops for human consumption also increased both the output and the productivity of arable land when lower-yielding crops were replaced by higher-yielding crops. The two major changes were the decline of rye and the appearance of the potato. The modern potato (*Solanum tuberosum*) was introduced into England in the late sixteenth century, and had been preceded by the sweet potato (*Ipomoea batatas*). For most of the following century it remained a curiosity and a luxury food, but by the close of the seventeenth century it seems that potatoes were fairly widely grown in the north-west as food for ordinary people. The physical environment here was suited to potato cultivation, and field systems were sufficiently flexible to accommodate the crop. Moreover, diets in this area were dominated by oats, and potatoes were a more attractive alternative to this crop than they were to wheat. A major growth in potato cultivation took place during the last quarter of the eighteenth century. Not only was population growing at an unprecedented rate during this period, but bad harvests, particularly those for 1794 and 1795 (see the prices in Figure 3.2), stimulated the cultivation of alternatives to cereals. As a correspondent to *The Times* put it in September 1795, 'From the apprehension of a second year of scarcity, potatoes have been everywhere planted and their produce has been generally great.' At the start of the new century the 1801 crop return suggests that only around 2 per cent of the arable area of England (excluding fallow) was under potatoes, although this proportion excludes potatoes grown in gardens and allotments. However, the returns show that for individual parishes the proportions could be much higher: 25 per cent in the mining parish of St Just in the far west of Cornwall for example. A further stimulus to the growth of the crop came during the Napoleonic wars with the very poor harvests of 1799, 1800, 1809, 1810, 1811 and 1812. As the century progressed potatoes became a food of those working on the land as well as those working in industry. Much of the new cultivation took place in small plots of land cultivated by agricultural labourers, in cottage gardens, in allotments, and in potato-patches in the corners of farmers' fields, and it has been estimated that by the middle of the nineteenth century potatoes from these sources were sufficient to provide over a pound per person per day for those working on the land. If modern evidence of the kilocalories available in wheat and potatoes (340 and 83 per 100 grammes respectively) is multiplied by the yields of the two crops in the early nineteenth century (around 500 kilograms and 5 tonnes per acre respectively) the result suggests that an acre of potatoes provided about two and a half times as many calories as an acre of wheat. National quantitative information on potatoes is not subsequently available until 1866; and by 1871 most counties had less than 5 per cent of their arable acreage (calculated on the same basis as the 1801 returns) under potatoes: the exceptions were Lancashire (25 per cent), Cheshire (20 per cent), Middlesex (11 per cent), West Yorkshire (9 per cent) and Cumberland (8 per cent).

The decline of rye is charted in Table 3.12. The crop had never been common in some counties such as Cornwall and Kent, but in others, especially Hertfordshire, Norfolk, Suffolk and Worcestershire, the proportion of the sown acreage under rye declines dramatically from the sixteenth century onwards. There is also evidence that the proportion of rye declined in other counties, including Leicestershire, Oxfordshire, and counties in the north and west. The replacement of rye by wheat implies that soils were improved and that calorific output per acre rose, but the change could also indicate a growing commercialisation of production since wheat was a cash crop and rye a subsistence crop. If rye were replaced by barley or oats and wheat bought in as bread corn, it would suggest that the decline of rye was related to regional specialisation allied to increased market integration.

Regional specialisation altered the distribution of crops and stock between regions. The advantages to be gained by specialisation are demonstrated in Table 3.16, which illustrates the principle of absolute advantage using hypothetical data. Two independent regions, A and B, need to produce 100 units each of two products X and Y. The yields of the two products differ (because of differences in the natural environment between the two regions) so differing resources (mainly land) have to be devoted to the two products. However, if each region devotes all its resources (15 units) to the commodity for which it has the highest yields, output across the two regions rises from a total of 400 to 600 units. As long as the costs of transporting the surplus of X from B to A, and Y from A to B are less than 200 units then it would be economic for each region to specialise. It is noticeable that when specialisation occurs, yields for the individual products (say QC/TC or QA/T for each region) do not change, but overall output for the two regions together rises. Since land input stays constant land productivity for the two regions (Q/T) will rise. Table 3.16 illustrates the simple case of absolute advantage; the situation of comparative advantage is more complicated but has similar effects.

Changes in the degree of agricultural specialisation have been measured indirectly by observing changes in the patterns of marriage seasonality recorded in parish registers. Kussmaul considers that a predominance of autumn marriages, taking place after the harvest at the start of the arable farming year, indicates a predominance of arable farming, and a peak of marriages in the spring is indicative of pastoral farming. For the period 1561–1640 the two types are fairly evenly scattered across the country, but by 1661–1740 distinct regional patterns emerge with an arable east (a peak of autumn marriages) and a pastoral west (a peak of spring marriages). This suggests a degree of regional specialisation, and implies an increase in trade. The difficulty with these findings lies in the relationship between marriage

Table 3.16 *The theoretical benefits of absolute advantage*

(a) Independent production

	Product X			Product Y			
Region	Yield	Resource units	Output	Yield	Resource units	Output	Total A + B
A	10	10	100	20	5	100	200
B	20	5	100	10	10	100	200
Total			200			200	400

(b) Specialisation and integration

	Product X			Product Y			
Region	Yield	Resource units	Output	Yield	Resource units	Output	Total A + B
A	10	0	0	20	15	300	300
B	20	15	300	10	0	0	0
Total			300			300	600

seasonality and farming type. An autumn marriage peak is assumed to indicate arable farming because people marry after the harvest at the end of their annual hiring contract as servants. For the spring marriage peak the assumption is that couples marry once calving and lambing are over. But this kind of animal husbandry is characteristic of rearing rather than fattening and so the apparent split between arable and pasture is really a split between livestock rearing and the rest.

However, confirmation of increased regional specialisation in the seventeenth century is provided by direct evidence from probate inventories. In eastern Norfolk, for example, an intensive mixed husbandry developed, centred on the production of wheat and the stall-feeding of bullocks with barley. East Worcestershire also saw a swing to wheat, but within a less intensive husbandry system which saw a reduction in livestock densities. The rise of a specialist dairying industry has been charted in Shropshire and Hertfordshire; in the Midlands there appears to have been a swing to permanent pasture for the fattening of cattle. Regional specialisation based on absolute or comparative advantage was recognised by one of the most astute observers of English agriculture, the Swedish botanist Pehr Kalm, who visited England in 1748.

In England the wholesome custom is much in use, that nearly every district lays itself out for something particular in Rural Economy, to cultivate, viz., that which will thrive and develop there best, and leaves the rest to other places. They believe they win more by this means than if they cultivated all departments of Rural

Economy; for, besides that he who has many irons in the fire must necessarily burn some, they also think it is not worth while to force nature ... They thus sell their own ware, and buy what they themselves have not, or they also exchange ware for ware.

Specialisation is much more evident by the mid-nineteenth century. By this time distinctions were both more varied and more subtle compared with earlier periods, as farmers were able to respond to a sophisticated marketing network.

Increases in crop output per sown acre

The ways in which agricultural output increased discussed hitherto have not included specific improvements in output per sown acre for individual crops. They have, however, involved changes in land productivity. For example, if the agricultural area of the country (T) is regarded as fixed, then all increases in output result in an increase in land productivity. A reduction in fallow also results in an increase in arable productivity but not necessarily in crop output per sown acre. Despite the many ways in which output and land productivity may rise, historians have tended to concentrate their attention on changes to crop output per sown acre, or yields per acre. This is partly because this index is the one most commonly used by farmers in the past (in terms of bushels per acre) and therefore is the most easily available index of land productivity. It is also important because, with the benefit of hindsight, crop yields per acre have shown the most potential for increasing output as a whole. Modern wheat varieties grown with plentiful supplies of nitrogen fertiliser can easily give yields of over 4 tonnes per acre (in 1800 wheat yields of 20 bushels per acre were equal to about half a tonne an acre). These potential gains far outweigh gains in output due to intensification or changes in the crop and livestock mix.

The simplest way to increase output per acre is to increase inputs. Under certain conditions, for example, higher seeding rates might increase yields, but it would be unlikely. The increased application of manure would be much more likely to increase yields, but even so there is a limit beyond which additional applications of manure will have no effect on yields and might even cause them to fall: too much nitrogen will weaken straw and cause a cereal crop to 'lodge' or fall over. Perhaps most effective of all would be the use of extra labour, especially for weeding. While these strategies could have been important for raising yields in the short term, in the longer term they are of less significance because they raise land productivities at the expense of other productivities. Most importantly the addition of extra labour is likely to have raised land productivity at the expense of labour productivity so that the overall efficiency of agricultural production may have been reduced. Moreover, increasing inputs in this way simply involves more of the same; without technological change the opportunities for sustained increases in land productivity are limited.

In order to understand how crop yields per acre might have been raised we need to know the main determinants of crop output per acre in the early modern period. Of obvious importance was the supply of crop nutrients and it is undoubtedly the case that increases in output per acre were associated with increases in the availability of such nutrients to cereal crops. This involves more than the assumptions of Figure 1.1, that such nutrients are a simple function of the amount of manure, and it is necessary to examine the supply of nutrients to the soil and the ability of growing crops to make use of them. Before discussing this, however, there are two other factors to consider: the incidence of pests and diseases and the development of new cereal varieties.

It is estimated that over one fifth of the world's cereal crop is currently lost to pests and diseases and it is obvious that such losses must also have been considerable in the past, probably exceeding this proportion. Methods of controlling, or attempting to control, pests, diseases and the pathogens responsible for spreading disease have been discussed in the previous chapter. The increase in the availability of agricultural labour in the sixteenth century would have contributed to the reduction in damage from some pests, by scaring birds for example, or by taking greater care of grain in store. Sometimes ingenious remedies were sought. Thomas Coke is reported by Arthur Young to have saved £60 worth of turnips by sending 400 ducks into the field to devour black canker caterpillars. 'In five days they cleared the whole most completely, marching ... through the field on the hunt, eyeing the leaves on both sides with great care to devour everyone they could see.' However, the most important way of controlling the build up of crop pests, and also of controlling disease, was through crop rotation, particularly those rotations that limited the number of corn crops taken in succession. New crop rotations evident in some areas from the 1720s and 30s, and more generally by the 1780s and 90s, were based on the principle that two corn crops should not be taken in succession, which, among other benefits, could limit the carry-over of disease. By the middle of the century such rotations were quite common and were imposed by landlords on their tenants through covenants in leases.

It is almost impossible to measure the impact of such developments. It is equally difficult to determine the importance of new crop varieties, which are one of the most important determinants of crop yield today. Scientific plant breeding is a twentieth-century phenomenon, but farmers in the early modern period were certainly aware of the benefits of selecting seed. Contemporaries described several types of wheat: John Houghton devoted several issues of his *Collections* to the description of wheat and quotes Robert Plot who described thirteen varieties being grown in Oxford in the late seventeenth century. The notion that certain varieties were more suited to certain types of land seems to have been well established, as was the principle that seed should be brought onto the farm from elsewhere. Hartlib for example considered that, 'it is excellent husbandry every year to change the *species* of Graine, and also to buy your Seed-Corne, from places farre distant'. Random mutations must have produced more productive varieties of cereal crops and it is likely that farmers would have selected seed from these in preference to others. The development of Chevallier barley by this process of selection in the early nineteenth century is described by Pusey. What impact this had on cereal yields remains an open question, since we have no evidence of varieties grown. In any case new wheat varieties may have been chosen for the quality of their flour rather than for their yield: indeed varieties with better quality flour for improved bread might have given lower yields per acre.

Perhaps the most important means of raising crop yields was by making more nutrients available to growing crops. The three principal plant nutrients are nitrogen, phosphorus and potassium, although these elements have to be available as chemical salts before they can be absorbed by plants. Before about 1830 it is fairly certain that nitrogen was the 'limiting factor' in determining levels of crop growth. Although contemporary farmers were unaware of the biochemistry involved, there were a number of ways in which more nitrogen could be made available to growing crops: by exploiting existing stores of nitrogen, by making more nitrogen already in the soil available to the plant, by conserving nitrogen supplies, and by adding new supplies of nitrogen to the soil.

The greatest store of nitrogen available for cereal crops was in the soils under permanent pasture. The easiest way to release this nitrogen was simply to plough up the pastures and grow cereals. Under modern conditions ploughed out grassland gives enough nitrogen for up to six years of wheat crop, but in the eighteenth century it may well have lasted for much longer – say up to twenty years. Nitrogen supplies regenerate once arable land reverts to permanent pasture and the nitrogen content of old arable land doubles every 100 years. The swings in the general balance between arable and pasture since the sixteenth century have already been discussed: the periods when most permanent pasture was ploughed up were in the late sixteenth and early seventeenth centuries, and again in the century after 1750. In the earlier period the nitrogen was exploited in land that had been converted to pasture in the fourteenth century. In the later period the conversion of pasture to arable reached its peak. The conversions were not necessarily permanent, and frequently the land would revert back to pasture, or be incorporated into a flexible rotation structure involving ley grasses, but considerable additions were made to cereal output, and initially at least to cereal yields. In the longer term, however, continued arable cultivation of pasture land may well have led to lower cereal yields. Table 3.6

shows that between 1800 and 1850 the acreage of meadow and pasture fell by 30 per cent while arable rose by 33 per cent.

As well as exploiting the nitrogen under pasture, more crop nutrients could be made available to plants by removing competition from weeds. Growing weeds could be controlled by hand-weeding, but this was a labour-intensive occupation and after a point crops could be damaged by the weeding process. Weeds were also controlled by better cultivation of the soil through ploughing and harrowing, which also encourages better root systems so that plants can take up more nitrogen. The replacement of fallows has been discussed as a way of increasing output per unit of arable, but there is also evidence that the remaining bare fallows were being cultivated more carefully, which would have had an effect on the yield of cereal crops following the fallow. Evidence from probate inventories, for example, indicates that as the seventeenth century progressed farmers in Hertfordshire were ploughing bare fallows more frequently. Inventories also indicate the spread of iron-shod ploughs, which would have been more efficient in the preparation of the land, and contemporary commentators also mentioned the virtues of deep ploughing, and discussed the types of plough most suitable for such activity. Such improved cultivation techniques were also necessary for the successful cultivation of turnips and clover, which require a finer tilth than do cereal crops.

Another way to make more nitrogen available to crops is to increase the rate at which organic nitrogen decays into mineral nitrogen which can be taken up by plants. The micro-organisms in the soil responsible for this require warmth, oxygen, water, and a moderate acidity. Improving these conditions will result in the conversion of more organic nitrogen to mineral nitrogen. Reducing soil acidity through the application of lime, for example, could produce a sudden spurt in nitrogen mineralisation. Farmers were well aware of the benefits of adding lime to the soil, as burnt lime, and later, as ground lime. The correction of soil acidity became particularly important with the spread of turnips which are particularly intolerant of acid soil. Marl was another substance frequently added to the soil. It was a mixture of clay and calcium carbonate and was much used both to improve soil structure and to reduce acidity. Although it had been applied to the land for centuries, the practice of marling became more widespread from the mid-seventeenth century. Marling made an important contribution to the reclamation of the Norfolk heathlands and is evidenced today by former marl-pits.

Lime and marl were important for improving soil structure, as were the improved cultivation methods already discussed, but soil drainage was the most important way of improving soil structure. Successful underdraining on a large scale had to wait until the nineteenth century with the introduction of the tile drain. Before then ridge and furrow was the principal means of surface drainage, but from the seventeenth century onwards hollow drains seem to have been more frequently employed, whereby stones or bushes were put into trenches and covered with soil. It is likely that the effectiveness of underdraining before the advent of tile drains in the mid-nineteenth century has been underestimated, since there are examples from the Midlands and East Anglia of quite dramatic increases in crop yields following underdraining in the late eighteenth and early nineteenth centuries. Nevertheless, until the mid-nineteenth century much of the heavy land in Figure 2.10 was subject to seasonal waterlogging, and as late as the 1870s it is estimated that over half the cultivated acreage needed draining.

These strategies had the effect of making more of the existing nitrogen available. More nitrogen could be added to the soil if existing stocks were conserved by managing supplies of manure more effectively. If livestock grazed permanent pasture then their manure would fertilise the grass but would have no effect on arable land. In some sheep-corn systems the sheep grazed the pasture during the day and were folded on the arable at night, but the most important way of conserving nitrogen stocks was to integrate grass and grain in rotations. Most efficient of all was the stall-feeding of livestock, particularly cattle, so that their manure could be collected and deposited exactly where it was needed. Stall-fed bullocks were not unknown in the seventeenth century, but it was not until the widespread cultivation of fodder crops, especially root crops, that the practice became common. By the third decade of the nineteenth century such 'high-feeding' was commonplace.

The nitrogen in fodder crops was usually kept on the farm and recycled through livestock. The nitrogen in cereal crops (excluding that in straw) was sold off the farm and thus lost. Some of this could be reclaimed in less orthodox sources of manure. From the sixteenth century onwards agricultural writers enthused about the possible sources of additional manure: seaweed, putrefying fish, silt, crushed bones, rags, malt dust, ashes and soot were all advocated. There is no way of knowing the extent to which these more esoteric substances were used, although it was likely to have been minimal except where a farm had good access to water transport. An obvious source of manure, from human beings, was little used before the nineteenth century, partly because of the difficulties of collection, but also because it was believed that it tainted crops. As rivers were improved and canals constructed in the eighteenth century, opportunities for importing manures from off the farm increased.

Despite the more efficient use of existing stocks of nitrogen, sustained increases in cereal output per acre were only possible with new additions of nitrogen to the soil. The most important source of new nitrogen was from leguminous crops which have a symbiotic association with nitrogen-fixing bacteria in their root nodules. Peas, beans and vetches had been cultivated since the middle ages, and, as Table 3.12 indicates, their acreages could be quite significant. Farmers had long recognised the value of legumes, as Tusser wrote in the sixteenth century,

> Where peason ye had and a fallow thereon.
> sowe wheat ye may well without doong thereupon.

The introduction of new legumes, especially clover, from the seventeenth century dramatically improved the amount of nitrogen fixed from the air. Clover fixes more nitrogen than peas or beans (some 55–600 kilograms per hectare per year compared with 33–160 kilograms for peas), and would remain in the ground for longer periods than pulse crops in arable rotations. For northern Europe it has been estimated that the introduction of new leguminous crops like clover increased the total nitrogen supply by around 60 per cent. This estimate is based on the assumption that legumes comprised 19 per cent of the arable area: in England the proportion was higher, reaching 26 per cent in 1871, so the increase in the nitrogen supply may have been proportionately greater.

Various clovers are indigenous to England, and probably formed part of natural grassland in some parts of the country. The introduction of sown clover leys is, however, a seventeenth-century phenomenon: as early as the 1620s there is evidence of clover seed being imported from the Low Countries. Slightly later, contemporary writers such as Hartlib, Yarranton, Blith and Worlidge enthused about the crop, pointing to its cultivation in the Low Countries as evidence of its value. These writers had no knowledge of nitrogen, or of how clover was a beneficial crop, but they did recognise that cereal crops following clover would benefit; as Blith put it: 'after the three or four first years of Clovering, it will so frame the earth, that it will be very fit to Corn again, which will be a very great advantage'. The first direct evidence of farmers sowing clover comes from the mid-seventeenth century, and the crop advanced on a wide front across the country. Other leguminous grasses were also grown, especially sainfoin on thin chalky soils. The evidence we have from Norfolk and Suffolk for the early eighteenth century suggests clover was not yet grown on a wide scale, since it accounted for only 3 per cent of the arable acreage, although, as Figure 3.9 shows, some 20 per cent of farmers were growing the crop. From the thirteenth to the early eighteenth century the proportion of the arable area sown with legumes in Norfolk ranged between 9 and 14 per cent; by the second quarter of the nineteenth century it was over 25 per cent. By the 1830s, when the first nationwide statistics for clover and 'seeds' are available (Table 3.14), it was accounting for over 30 per cent of the arable acreage in some counties. Thus the chronology of the spread of clover is similar to that of turnips. It was being grown in the seventeenth century but had relatively little impact on husbandry systems until after 1750.

The addition of nitrogen from leguminous plants like clover probably meant that, by the second quarter of the nineteenth century, nitrogen was no longer the limiting factor in the growth of cereals. By this time it is likely that phosphorus was the limiting factor and from the 1830s the amount of phosphate added to the soil began to increase dramatically. Most of these phosphates were imported, especially in the form of bones. The German agricultural chemist Liebig considered England was 'robbing all other countries of their fertility ... in her eagerness for bones'. Bonemeal is a slow-acting fertiliser and an important breakthrough took place when Lawes pioneered the process of adding sulphuric acid to bones to make superphosphates in the 1840s.

Increases in livestock productivity

Evidence about the productivity of crops is much more plentiful than evidence about the productivity of livestock, which is unfortunate, since the magnitude of changes in livestock productivity appears to have been greater than that for crops. Livestock productivity could have risen through two processes; first through an increase in the number of animals supported by a given area of land (A/T or A/TFOD) brought about by an increase in fodder supplies, and second, because livestock became more efficient at converting fodder into saleable livestock products (QA/QFOD). These two aspects of livestock productivity will be considered separately: first the improvement in densities, and second the improvement in animals.

The total number of animals per unit of farmland (A/T) can increase simply by switching from growing cereals for humans to growing fodder for animals, but the overall effect of this would be to lower total output since cereals yield more human food per acre than do fodder crops. Thus the most important changes in fodder supply are those which result in an increase in fodder output per acre (QFOD/TFOD). The earliest date possible for the measurement of livestock densities (A/T or A/TFOD, for example) is in the 1830s when both crop and livestock statistics are available for a few counties. For earlier periods probate inventories list the number of animals but only record the acreage under sown crops, excluding fallow, meadow and pasture. In the one study available, for Norfolk, it was found that livestock densities (A/TC) doubled in the first half of the seventeenth century. This coincides with the period for which evidence exists for improvements in fodder output per acre. In the early sixteenth century pastures and meadows were in a natural biological state, but by the end of the seventeenth century many were systematically cultivated, fertilised and sown with seeds imported onto the farm. This heralded the abandonment of the division between permanent arable and permanent pasture (discussed further below) and increased the output of grass per acre. As the seven-

Table 3.17 *Evidence of 'grassland' farming in Norfolk and Suffolk,* *1584–1739*

Period	% farms with 'summerleys'	% farms with 'grass'
1584–1599	0.0	13.7
1628–1640	0.0	22.5
1660–1699	2.5	22.8
1700–1739	8.8	26.5

teenth century progressed, clover became increasingly incorporated in the sown leys. Clover is particularly suited to areas with little natural pasture or meadow, and especially where natural grasses would take some time to establish themselves in a temporary ley. Table 3.17 shows how sown grasses were becoming more important in Norfolk and Suffolk. It is based on probate inventories which were supposed to exclude crops grown without 'the industry and manurance of man'. Thus the increase in the recording of grass and summerleys suggests that they were not natural, but had been sown on prepared ground.

The seventeenth century also witnessed the improvement of meadows through 'floating'. The floating of watermeadows was the process whereby a thin film of river water was kept flowing over the grass during the winter. This moving water kept the meadow frost-free and encouraged the growth of early grass, providing fodder, usually for sheep, in March and April when fodder shortages were usually most acute. After the flock had been moved to summer pastures the meadows would be irrigated again and substantial hay crops taken in June or July. The floating of meadows was the development of a natural phenomenon, for as Speed pointed out in 1611 'rivers do so batten the ground, that the meadows even in the midst of winter grow green'. The water was kept flowing by two types of system. Simple catchwork systems involved channels cut along contours of a valley side with water flowing down the valley from one channel to the next. Bedwork systems were more sophisticated, involving channels on ridges and drains in the furrows.

The earliest documentary evidence of floating is from 1608 when watermeadows were referred to in the court rolls of Affpuddle in Dorset, and in 1629 a waterman was appointed for the whole manor. Floating was most common on valley floors in chalkland areas: on the Frome and Piddle in Dorset and on the Avon in Wiltshire. By 1700 most of the major rivers in Wessex (Dorset, Hampshire and Wiltshire) had meadow systems and the main phase of development was over by 1750. By the 1830s, it was estimated that there were 6,000 acres of watermeadow in Dorset, although another estimate for 1866 thought there were only 10,000 acres in the whole country. Watermeadows were much less common in the east of England: they were naturally more suited to the wetter west.

These improvements in the productivity of conventional fodder supplies from pastures and meadows (QG/TG) were supplemented from the eighteenth century with new fodder crops in arable rotations. These new sources of fodder, turnips and clover, increased fodder yields because the fallows and permanent pastures they replaced were much lower yielding sources. Thus the output of animal products per area of fodder (QA/TFOD) was increasing, as the composition of the fodder area (TFOD) changed, reducing the proportion of fallow (TFAL) and permanent grass (TG). There is also evidence, from descriptions of how turnips were cultivated, that turnip yields were improving, so that QAF/TAF was also increasing. All else being equal, as the output of fodder per acre increased, so would the number of animals and the output of animal products per acre.

The number of animals alone, however, is not an adequate guide to the output of livestock products because of the period of time animals take to mature. If the time taken for cattle to be ready for slaughter reduces from four years to two, then output can double while the stock of animals at any one point in time remains the same. We have no reliable evidence on this, only the opinion of Gregory King that, in the late seventeenth century, less than one fifth of the nation's cattle stock was slaughtered each year, whereas around the turn of the nineteenth century it was about a quarter, implying a 25 per cent improvement in the supply of animal products. Unfortunately the evidence is not available to demonstrate that increases in fodder supplies in the eighteenth century gave rise to these increased livestock densities.

Increased fodder supplies increased the output of animal products. These gains would have been augmented by changes to livestock which increased the rate at which fodder was converted into meat, wool and other products. Improvements in livestock themselves are also difficult to measure, despite the din of propaganda from a few very successful livestock breeders in the eighteenth century, and a growing volume of evidence of the rapid spread of new livestock types. Improvements to livestock by selective breeding was not a new phenomenon in the eighteenth century, since the principle of selecting animals for certain purposes is as old as domestication itself. The famous Robert Bakewell, for example, was an excellent publicist with a keen eye to the commercial main-chance: his major contribution was in formalising and publicising the methods of selective breeding. A recent study has argued that Bakewell's contribution to the development of Longhorn cattle in the eighteenth century was minimal; the breed was the product of traditional breeders using traditional methods operating over a time span of at least one hundred years.

The most successful new cattle breed was not the Longhorn but the Shorthorn. By the mid-nineteenth century over half the cattle in England were Shorthorns or Shorthorn crosses. The breed developed using the principles of close in-breeding applied to the cattle of County Durham and North Yorkshire. Several breeders were involved, of whom Thomas Bates and the Collings brothers were the most important, although the animal had been improved before then, possibly through the importation of Dutch cattle in the seventeenth century. The Shorthorn is a dual-purpose breed and was well suited to the demands of high farming in the nineteenth century, although doubt has recently been cast on the merits of the animal and the extent to which it entered into the national herd. Herd book evidence shows how the breed spread from the diffusion hearth on the borders of Durham and North Yorkshire in the 1770s to the rest of the country by the mid-nineteenth century.

Distinctive new sheep breeds began to develop towards the end of the eighteenth century. The objectives of the breeders were to redistribute flesh to the expensive parts of the animal, to reduce the proportion of bone and offal, to improve conversion of food to meat, and to get the animal to maturity as quickly as possible. From the mid-1740s Bakewell began experiments with sheep and by the 1790s there were fifteen or twenty breeders of Bakewell's calibre in the Midlands. Bakewell developed the longwool New Leicester sheep, which was important in its own right, but especially valuable when crossed with other breeds. The most important shortwool sheep, the Southdown, was established by John Ellman of Glynde in Sussex. These two breeds were the foundations of sheep breed improvements of the late eighteenth and early nineteenth centuries.

The spread of new breeds was closely related to changes in the role of sheep in farming systems, and especially to the fodder available for them. There is evidence from the south of England that ordinary farmers developed sheep breeds in response to the relative prices for meat, tallow and wool. During the eighteenth century the demand for large fatty joints of mutton and for tallow led to an emphasis on the horned breeds such as the New Leicester. In the 1790s attention turned to the Southdown which gave high-quality wool; but after 1815 the Southdown and other new Down breeds were developed to take advantage of new arable systems with the intensive cultivation of fodder crops, meeting a growing demand for good-quality mutton. This was also the case in Oxfordshire, where the New Leicester was associated with enclosure and the cultivation of turnips. Local breeds were crossed with the New Leicester: the improved Cotswold was kept by 5 per cent of farmers in 1820 and 25 per cent by 1840; only 1 per cent of farmers had the Oxford Down in 1830, but some 45 per cent by 1850. In Norfolk and Suffolk, the rather unruly indigenous Norfolk, valued for its wool and role in the foldcourse, began to decline in the early nineteenth century, to be replaced by the more amiable Southdown (and later with the Norfolk–Southdown cross which became known as the Suffolk), kept primarily for its ability to turn turnips into mutton. What impact these breed changes had is hard to observe directly. However, estimates of the volume of output of animal products show a two-and-a-half-fold increase between 1700 and 1850, yet the stock of animals in the country hardly seems to have increased at all. This suggests considerable improvement in the productivity of livestock (in terms of output per animal). This was partly due to improvements in fodder, partly due to breed changes and partly due to an increased turnover of animals.

There is no direct evidence of livestock weights or yields of livestock produce which could be used to measure output per animal before the nineteenth century. However, we do have information on the prices of livestock and the prices of their products (meat and wool), and the ratio between the two might be indicative of output per animal (the price of cattle divided by the price of beef per pound for example should give some indication of the number of pounds of beef per animal). This information is shown in Table 3.18 and Figure 3.7(c), but unfortunately (and rather surprisingly) there are no livestock price series for the century after 1760. For what they are worth, the price-ratios indicate no change in the productivity of cattle between the mid-sixteenth and the mid-eighteenth centuries, but an increase for both mutton and wool of some 78 per cent during the first half of the eighteenth century, in comparison with the preceding century. It is also evident that the price of pigs relative to cattle had been increasing continuously since the sixteenth century; in the second half of the sixteenth century seven pigs were equivalent to one cow, by the mid-eighteenth century the number had reduced to three.

The absence of improvement in the size of cattle is confirmed by archaeozoological evidence which suggests that the increase in the size of cattle took place between the middle ages and the sixteenth century, rather than later. If the broad indications of trends in output per animal for sheep are correct, then improvements in wool and mutton yields took place before the breed developments of the late eighteenth century, and therefore presumably must reflect improvements in fodder supplies and in the management of sheep flocks. This evidence corroborates recent opinion that sheep were improving in size from the mid-seventeenth century as a consequence of improvements in the supply of fodder. Yet the lack of change in the ratio of beef and cattle prices would suggest that no significant improvements in fodder supplies were having an effect here. The rise in the value of pigs relative to cattle suggests that pigs were increasing in size. Pigs respond well to increased feeding and can eat a great range of foods. They are particularly useful for clearing land and the increase in their size and value could therefore be associated initially with the more

Table 3.18 *Ratios of animal and animal product prices, 1550–1750 (index numbers, 1550–1750 = 100)*

	Cattle/ Beef	Sheep/ Mutton	Sheep/ Wool	Swine/ Cattle	Horses/ Cattle	Sheep/ Cattle
1550s	–	–	65	65	92	95
1560s	88	–	76	63	90	99
1570s	116	–	87	60	83	95
1580s	111	–	89	63	82	90
1590s	107	–	90	61	83	106
1600s	98	76	75	68	96	98
1610s	109	80	83	74	84	88
1620s	117	78	83	70	84	82
1630s	114	82	79	94	97	86
1640s	106	90	99	95	100	96
1650s	105	69	69	98	99	82
1660s	92	65	64	113	112	90
1670s	98	76	77	106	103	92
1680s	92	80	98	130	116	105
1690s	113	91	87	118	95	96
1700s	98	107	101	136	106	111
1710s	87	106	116	128	115	102
1720s	85	117	137	132	120	106
1730s	76	135	143	171	112	117
1740s	82	187	201	155	125	152
1750s	105	162	180	–	106	113

intensive use of waste land. Later, as the areas in which they foraged were put under the plough, pigs might have been kept indoors and fed more intensively.

The integration of grass and grain

So far the various contributions to improvements in productivity, and the cropping changes associated with extensions of the arable area, have been considered separately, but the whole was greater than the sum of the parts, and as these various changes were introduced they evolved into new systems of farming. The two systems which are most important are convertible husbandry and the Norfolk four-course rotation. Convertible husbandry is the system where the distinction between permanent grass and permanent arable is broken; arable land rotates around the farm. At its simplest, permanent pasture was broken up and cropped with corn for a few years, and then the land was allowed to revert to grass for some time, perhaps over twenty years, but more sophisticated systems of convertible husbandry would have much shorter grass leys of a year or two. Kerridge has made much of convertible husbandry, which he calls 'up-and-down' husbandry, and it forms the cornerstone of his 'agricultural revolution'. The main period of its spread was between 1590 and 1660. Other historians consider the impact of convertible husbandry farming on yields would have been minimal. If permanent pasture were ploughed in the course of the introduction of convertible husbandry, the store of nitrogen released could have had a dramatic short-term influence on the yield of cereal crops. Nevertheless, within a period of a few years, yields would have fallen back to their previous levels as the amount of organic matter decreased, and the soil became more acid because of leaching and the production of acids from the decay of organic matter. Thus the development of convertible husbandry from the mid-sixteenth century could be interpreted as a means of cashing in on reserves of nitrogen under permanent pasture for short-term gain. Indeed, there is some evidence of a retreat from 'up and down' husbandry in the Midlands in the later seventeenth century once these gains had been made and yields were probably starting to fall. Leys could also have been introduced without ploughing up pasture by sowing grass on the fallow field. In the late sixteenth century, for example, about 20 per cent of the open arable fields of Wigston Magna near Leicester were described as 'grass ground'. Even so, it was also difficult to establish a grass ley: 'to make a pasture breaks a man, to break a pasture makes a man'. But once established a grazed grass ley would serve to keep down weeds, leave organic residues when it was ploughed in, and serve as a barrier to the transmission of some crop-specific pests and diseases. Moreover, the increased cultivation needed to produce leys could improve the soil structure.

The main advantage of convertible husbandry was that it integrated livestock into arable farming. This process was continued with the development of a rotation that came to be known as the Norfolk four-course, illustrated in Figure 3.10. Grain crops alternated with a fodder crop to reduce the incidence of pests and diseases, nitrogen was added by clover and subsequently recycled by livestock consuming the roots and seeds. Weeds were controlled if turnips were hoed. The rotation also had other advantages. Although it required more labour, especially for hoeing turnips, demand for that labour came in seasons of the year when labour demand was usually slack. Thus it did not affect labour requirements at peak periods and served to even out the demand for labour over the year.

The effects of the rotation were to increase yields of grain and to allow much higher stocking densities. These effects are modelled in Table 3.19 which demonstrates convincingly that the Norfolk four-course could indeed have been responsible for unprecedented changes in both crop and livestock productivity and output. The first model farm in the table (A) has 40 per cent of its area devoted to grass in the form of permanent pasture and Of

Table 3.19 *The impact of the Norfolk four-course rotation*

Model farm	Grain yields (bushels/acre)	Grain output (bushels)	Livestock output[a] (bushels)	Total output (bushels)	% grain
A	11.5	460	400	860	53
B	21.4	642	950	1492	43
	(23.9)	(717)	(950)	(1567)	46
C	16.0	800	750	1550	52
	(18.5)	(925)	(750)	(1675)	55

Figures in brackets assume some pest and disease control from the rotation.
[a] Assuming the rotation has all the benefits of fallowing
A 40% permanent grass; the remaining 60% under a three-course rotation of 20% wheat, 20% oats, and 20% fallow.
B 40% permanent grass; the remaining 60% under a four-course rotation of wheat 15%, turnips 15%, barley 15%, clover 15%
C No permanent grass; four-course rotation of wheat 25%, turnips 25%, barley 25%, clover 25%.

meadow and the remainder under a three-course arable rotation of wheat and oats followed by a fallow. The third (C) has the whole farm devoted to the Norfolk four-course, so that the rotation of wheat, turnips, barley and clover now covers the entire farm. A wholesale switch from one to the other would have been most unlikely in practice, at least in the early eighteenth century, because of the risks involved in replacing permanent pasture with turnips and clover. Permanent grass was a reliable if low-yielding source of fodder, but the new crops required new cultivation techniques and had a much higher risk of failing. Thus the second farm (B) retains the acreage of grass and the Norfolk four-course is implemented over the arable area previously under a three-course. Although this farm retains the assured source of fodder, the *area* of grain is reduced by 25 per cent (from 40 per cent to 30 per cent of the farm area). If a farmer moved from a three-course to a four-course system (farm A to farm B), retaining his area of permanent grass, grain yields would have doubled (a rise of 107 per cent) and total output would have risen by 82 per cent. Grain output would have increased despite the reduction in area because of the dramatic increase in yields, but the major proportion of the increase in total output would have come from the livestock sector. If a farm then moved towards system C and devoted an increasing proportion of its area to the four-course, yields would have started to fall but grain output would nevertheless have risen because of the additional area under grain. Furthermore, as the arable area encroached onto areas hitherto under pasture the initial rise in yields would have been higher than those indicated in the table because of the utilisation of nitrogen reserves in permanent pasture.

course the rotation was not without its problems. Turnips and clover required a fine tilth which was difficult to obtain on heavy soils. Turnips also required the land to be limed, were subject to club-root disease and the ravages of the turnip fly, and could be killed by a hard frost. In practice it was very difficult to grow clover every four years because the land became 'clover sick' and so the Norfolk four-course was rarely implemented year after year in this pure form, the most usual variant being the extension of the clover ley for another year or two before ploughing for wheat.

Some idea of the extent of the rotation is gained from Table 3.14 which shows the earliest available national statistics on the acreages under clover and seeds. As might be expected, the figures for Norfolk are remarkable in showing that wheat, turnips, barley and clover each accounted for about a quarter of the arable acreage in the 1830s. In other counties the proportions were lower, but still suggest the prevalence of the rotation. The first occurrence of the Norfolk four-course is more difficult to identify. Although both turnips and clover were quite common in Norfolk by 1750, acreages were small in relation to grain crops, and the Norfolk four-course was not widespread until after 1800. An early documented occurrence is on three

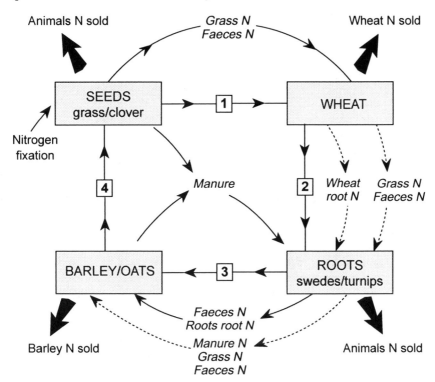

Figure 3.10   The Norfolk four-course rotation. *Source:* Shiel (1991), 68.

Table 3.20 *Norfolk: trends in agricultural production, 1250–1854*

	1250–1349	1350–1449	1584–1640	1660–1739	1836	1854
% Grain[a]						
Wheat	19	18	29	20	48	49
% Sown Area[b]						
Grain	87	87	87	84	49	52
Legumes	14	13	9	14	27	24
Clover	0	0	0	2	25	21
Turnips	0	0	0	7	24	22
Livestock ratio[c]	32	36	51	70		61
Draught beasts[d]	8	6	14	11		11
Grain Yields[e]						
Wheat	15	12	15	15	23	30
WACY[f]	11	9	9	10	21	26

[a] Percentage of wheat, rye, maslin, barley and oats
[b] Area sown with arable crops excluding fallow
[c] Livestock units per 100 cereal acres
[d] Oxen and horses per 100 sown acres
[e] Bushels per acre
[f] Weighted Aggregate Cereal Yield

Norfolk farms from 1739–51, and at about the same time on a farm just
north of Ipswich. Table 3.20 illustrates both the nature and chronology of
this new farming for the county of Norfolk. Medieval data is combined with
the early modern to give a time perspective of six centuries. What is striking
is the stability of the pattern for the 500 years from 1250, and the dramatic
changes that take place in the succeeding century.

Partly because these integrated mixed-farming systems comprised so
many mutually dependent components their evolution took time. Hence the
long lag between the appearance in England of clover, turnips and the other
components of the Norfolk four-course system and the perfection of the
system itself, whose widespread diffusion must be dated to the first half of
the nineteenth century. Nevertheless, there can be no doubt of the superi-
ority of the new system at whose root quite literally lay the improved
management of soil nitrogen. This of course, was not farmers' intention,
since the chemistry was unknown to them. Their concern was with fodder.
Sowing grass leys focussed attention on the range and suitability of grasses
that could be grown and so clover was selected as appropriate for a temporary grass; turnips were an alternative source of fodder. Once grown, and
integrated into arable rotations, their probably unintended outcome was an

increase in overall output. Although systems such as the Norfolk four-course rotation increased output of both stock and crops their major contribution may have been that optimum output occurred with a larger proportion of arable crops than under a system of permanent grass and permanent arable. The much-increased amounts of manure from more efficient fodder crops, and the rotational use of crop residues, allowed this substantial increase in grain area, while still maintaining, or even boosting, yields.

Labour productivity

Improvements in the productivity of labour in agriculture from the eighteenth century have often been portrayed as the 'release' of labour from agriculture to industry and therefore historians have looked for rural–urban migrations as the necessary evidence for productivity change. This over-complicates and confuses the matter. What is at issue is the *proportion* of the workforce in farming, not the absolute number; indeed the agricultural workforce can be increasing its size at the same time as labour productivity is rising. The estimates of labour productivity discussed earlier are extremely rough and ready. They suggest that the growth in labour productivity was very slow during the sixteenth century, at only 8 per cent over the eighty years from 1520 to 1600. By 1700, and probably from the mid-seventeenth century, labour productivity was showing a sustained rise. In the absence of any detailed studies of the determinants of labour productivity at local level we are left with a string of untested hypotheses as to why this was the case. Labour productivity will increase as land productivity increases, provided the latter can improve without additional labour. However, many of the contributions to rising land productivity made greater demands on agricultural labour. The replacement of permanent pastures and fallows by convertible husbandry involved more land preparation through ploughing and harrowing, and more time generally on managing crops. Turnips in particular, if cultivated correctly, were demanding of labour through hoeing and lifting. While some agricultural operations required the same labour input irrespective of crop yields, such as ground preparation, many were directly proportional to yield such as threshing and to a slightly lesser extent harvesting. Thus higher yields inevitably meant more labour was required unless harvesting or threshing technology changed. The list of possible explanations for improvements in labour productivity is long and growing, but it may be divided into four categories: mechanisation and changes in labour practices, improvements in the amount of energy available in farm work, increases in farm size, and changes in employment practices.

Little mechanisation of farming took place before the mid-nineteenth century, but before then there can be little doubt that small incremental improvements were made to basic farm implements over the centuries. The most important improvement before the mid-eighteenth century was the substitution of iron for wood for the parts of ploughs and harrows that were subject to the most wear. But there were also more general improvements in the design of ploughs. In the sixteenth century ploughs were generally wheeled and heavy. By 1600 the lighter, two-horse 'Dutch plough' was to be found in parts of eastern England. A dramatic improvement was the Rotherham plough patented by Disney Stanyforth and Joseph Foljambe in 1730. This was a light general-purpose swing plough (that is with no wheels) that was easy to make, cheap to produce and yet stronger than other contemporary ploughs. Fewer horses were needed to pull it, and there was consequently less need for a man or boy to tend the horses; ploughing could be carried out by one man. It appears that the plough was made from standard patterns and that all parts were interchangeable. Advertisements for the plough claimed it reduced ploughing times by a third, or required one third less horsepower. The plough was adapted in Norfolk by Arbuthnot by the time of Young's *Eastern Tour* in the late 1760s. Improvements multiplied towards the end of the eighteenth century, undertaken by men such as Ransome of Suffolk, the most important being the replacement of wood by cast iron, so that by the turn of the century ploughs were being made in rural foundries rather than by the local blacksmith.

Another eighteenth-century development was the seed drill and the horse hoe. Jethro Tull is familiar as inventor of the seed drill in 1731; in fact designs for seed drills were published in the early sixteenth century. Yet it was over forty years after Tull's invention before it attracted imitators and 120 years before the drill was widespread in English agriculture. Tull's method was to sow seed in rows with a drill and hoe between them with a small plough (the horse hoe). This used only 30 per cent of the normal seed requirement (3 pecks per acre for wheat instead of 2.5 bushels) since the rows were so far apart. By the time of the Board of Agriculture Reports at the turn of the nineteenth century, drilling had become common in Northumberland and Durham, and in Norfolk and Suffolk, but the use of the horse hoe was less common. Figure 3.11 provides some evidence for the spread of both drill and horse hoe for Oxfordshire and the Welsh borderland using the evidence of farm sale advertisements. Drills appear in small numbers from the 1770s, but it is not until after 1810 that diffusion is rapid, and not until after 1850 that horse hoes begin to spread.

The first major change in harvesting technology was the shift from shearing with the serrated-edge sickle to reaping with a smooth-edged hook, then to 'bagging' with a heavy smooth hook, and finally to using a scythe. Until the mid-eighteenth century wheat was cut with a sickle,

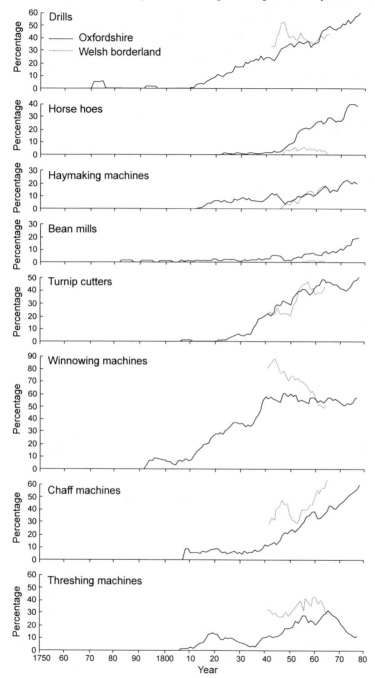

Figure 3.11   The diffusion of implements and machines in Oxfordshire and the Welsh borderland, 1750–1880. *Source:* Walton (1979), 25.

Table 3.21 *Work rates and labour requirements for hand tools and machines*

	Cutting (acres per day)	Worker days per acre including followers[a]
Wheat harvesting		
High reaping (sickle)	0.35	3.6
Low reaping (sickle)	0.25	4.8
Low reaping (reap-hook)	0.33	4.0
Bagging hook	1.0	3.0
Scythe	1.15	2.4
Manual delivery reaper	10.0	1.1
Self-delivery reaper	10.0	1.0
Reaper binder	10.0	0.5
Hoeing		
Hand hoe	0.3	3.3
Horse hoe	4.0	0.25
Wheat threshing		
	(Tons per day)	
Flail	0.2	5.0[b]
2-horse thresher[c]	5.0	1.6
12 hp steam thresher[d]	15.0	0.8

[a] Gathering, binding and raking
[b] For a 30 bushel crop
[c] Excluding winnowing
[d] Including winnowing and dressing

usually with a high cut in southern and eastern England, and a low cut elsewhere. North of a line from Chester to York, barley and oats were also reaped, but in the south they were mown with a scythe. The move to using a bagging hook and then a scythe with wheat began in southern England during the Napoleonic Wars (1799–1815), but it was not until the years after 1835 that the change was widespread: around 1790 some 90 per cent of the wheat harvest was carried out with the sickle; by 1870 it was 20 per cent. The implications of the change are illustrated in Table 3.21. Collins argues that the primary incentive for change was a shortage of labour, particularly migrant labour, exacerbated by the rise in yields from the 1830s which increased demand for harvest labour. Perkins argues that the scythe was adopted for corn harvesting on the uplands of Lincolnshire and the East Riding during the latter half of the eighteenth century as the rapid conversion of permanent pasture to tillage initiated a shortage of harvest labour. The mowing of grass and clover was already commonplace in these areas whereas the little reaping that was carried out was mostly done by women. The scythe saved grain because grain was less likely to shed, the grain was cut low and the short stubbles could be raked.

The first successful threshing machine was developed in Scotland and began to appear towards the end of the eighteenth century, coinciding with a shortage of labour during the Napoleonic Wars. The Board of Agriculture Reports give an indication of its spread: in the early 1790s into north-east England, and around 1808 into most counties of England, but with concentrations in the north-east and the south-west. These early machines were permanently installed in barns and were driven by horse power. Their use became much less common after the Wars, as the labour shortage abated, but they began to reappear in the 1840s and 50s (see Figure 3.11). By then they were much more substantial, usually mobile, and powered by steam. Their price had also fallen. Other, smaller, labour-saving machinery was also introduced including winnowing machines, turnip cutters, chaff cutters, bean mills, and rather later than these, oil cake-crushers. Reaping machines did not appear until the 1850s.

The 1830s were an important watershed in the history of farm mechanisation because it was not until then that the agricultural engineering industry developed. On the supply side, from the 1830s the price of iron fell and the development of machine tools reduced costs and produced a more reliable product. It is difficult to estimate what the total effect on labour productivity from the introduction of these new implements and machines might have been. Table 3.21 gives some indications, but probably exaggerates the savings that were achieved in practice. In harvesting and threshing wheat, machinery saved about 70 per cent of labour but the figure was lower for barley and oats. Over the course of the nineteenth century as a whole it has been estimated that labour requirements per acre in corn-growing fell by 30 per cent. But in other areas of farming the reduction was much less.

Machines saved labour because they enabled certain tasks to be carried out more effectively and because they made more efficient use of human energy or replaced human energy with animal or inanimate energy. Another important influence on labour productivity was the amount of energy available for farm work. A major source of energy was of course human labour power. During the period 1750–90 it has been suggested that nutritional standards for agricultural labourers fell in the south relative to the north of the country, where wages were higher. But in both cases the daily intake of kilocalories averaged 2,000, which is lower than that necessary to sustain hard physical work. This argument could be applied chronologically as well as spatially. Thus the apparent upsurge in labour productivity from the mid-seventeenth century might be linked to the relatively low price of foodstuffs and the rise in real wages, especially during the first three decades of the eighteenth century.

There were also important changes in the nature and availability of animal power in agriculture. Horses had been replacing oxen since the middle ages as the main form of traction on farms. Horses had an advan-

Table 3.22 *Horses and labourers in English and Welsh agriculture,*
1700–1850

	Horses (m.)	Arable (m. acres)	(1)	(2)	(3)
1700	0.50	9.00	5.58	0.90	2.79
1800	0.70	11.50	6.09	1.00	3.75
1850	1.00	15.30	6.54	1.10	4.55

(1) Horses per 100 arable acres
(2) Adult males employed in agriculture (m.)
(3) 'Manhours' of horse power available per man

tage over oxen in terms of both speed and strength: they could work 1.5
times faster than oxen and, in theory, could reduce labour requirements by
a third. Already by the fifteenth century some counties were more reliant on
horses than they were on oxen, particularly in the east of the country; in
Norfolk, for example, over 70 per cent of draught power was provided by
horses, in Essex 68 per cent and in Suffolk 51 per cent. The figure was much
lower in other parts of England: 6 per cent in Somerset, 15 per cent in
Durham and 18 per cent in Sussex, for example. This work on the middle
ages has not been extended (except for Norfolk where oxen had virtually
disappeared by the 1630s), but it is clear from contemporary commentators
that oxen had been replaced in most areas by the mid-eighteenth century.
A shortage of horses for farm work seems to have ensued after the
Napoleonic Wars, and the ox enjoyed a brief resurgence.

The substitution of animal for human labour and effort has recently been
suggested by Wrigley as another potential source of rising labour productivity. He shows that *pro rata* English farmers had two-thirds as much
animal power at their disposal as their French counterparts at the turn of
the nineteenth century. Table 3.22 gives some estimates of the number of
horses in agriculture together with estimates of the size of the male labour
force. A doubling in the number of horses from 1700–1850 coincided with
the expansion of the arable acreage, but even so there was a 17 per cent rise
in the number of horses per arable acre. The growth in the size of the labour
force was much less rapid, so the amount of horsepower available to each
worker grew by 34 per cent (0.30 per cent per annum) between 1700 and
1800; from 1800 to 1850 it was 21 per cent (0.39 per cent per annum). This
is expressed in the table in terms of 'manhours', on the assumption that one
horse did the work of five men, to demonstrate how human labour could
be saved, and so labour productivity increase.

Animal power was used for purposes other than hauling the plough and
harrow and pulling the cart or waggon. Animate energy had long been used

to grind corn, using a horse mill, or gingang as it was known in the north-east. This was powered by a horse walking round in circles, harnessed to a beam, which transmitted the rotary motion to mill stones through a system of cogs. From the late eighteenth century, horse mills were used to power an increasing variety of barn machinery mentioned above. It was not until 1840 that an efficient portable steam engine adapted for agricultural purposes first appeared. From the 1850s traction engines were being used for tasks on the farm, most commonly for threshing, where they moved from farm to farm, although the larger farms would have a stationary steam engine installed to power the thresher and other barn machinery.

Farm size is linked to labour productivity simply because larger farms appear to have employed fewer people per acre, so that if average farm size increased, the average number of employees would decrease. The general relationship between the growth in farm size (discussed in the next chapter) and the improvement in labour productivity is evident, as is the relation-ship between labour productivity growth and enclosure. Allen argues that this was the case from the mid-seventeenth century, using more specific evidence on the relationship between employment and farm size from data compiled by Arthur Young in the 1760s, applied to a new body of data on farm size in the South Midlands. These farms were growing in size during the eighteenth century and by implication would have been using less labour per acre. Allen has calculated that half the growth in output per worker from 1600 to 1800 was due to increases in farm size. On the other hand, evidence from Belgium and Ireland in the nineteenth century suggests that small farms could be more efficient in their use of labour than larger ones. Increasing farm sizes could be associated with the process of enclo-sure discussed in the next chapter. It is possible that enclosure increased labour productivity if it replaced farms comprising scattered strips of land in the open-fields with new compact ring-fence farms. Contemporaries were convinced that scattered strips wasted labour time and were therefore inef-ficient, but no empirical studies are available of the possible increases in labour efficiency following enclosure.

Changes in employment practices cover many possible factors. Evidence is available of changes in employment relationships, but the effect of these on labour productivity (as opposed to other effects) has not really been explored. Some changes in employment practices from the eighteenth century could have increased labour productivity defined by using the proportion of the labour force as the denominator in the calculation, but would have made little difference to labour productivity calculated in terms of output per man hour. New fodder crops made increased demands on labour (particularly for hoeing and lifting turnips) but these demands came at seasons of the year at which the demand for labour had previously been slack. The introduction of new crops therefore served to even out the demand for labour throughout the year. This would not have required an increase in the number of people employed in agriculture, but would have meant those already employed working more hours over the year, although in those cases where labour on the farm was provided by family labour and by servants on an annual contract (as opposed to day labourers) this additional labour would have cost virtually nothing. The supply of labour was forced to fit more closely to the demand by reductions in the duration of many labour contracts: from the year, to the week, or sometimes to the day; often the process went further and workers were paid by piecework rather than a flat rate. Thus proportionately fewer agricultural workers would be needed. This accords with the decline in the incidence of farm servants, since servants were hired by the year on an annual contract.

Changes in farm management

Historians have naturally tended to concentrate on the more conspicuous aspects of agricultural change, such as the introduction of new crops and new implements. But efficiency or productivity gains could also be brought about by improvements in farm management, or, in simple terms, an improvement in the ability of farmers to farm. Modern studies of productivity in farming have found the practical and technical ability of the farmer together with good labour management to be very closely related to overall farm productivity, yet we have virtually no historical evidence of changes in the level of farming competence. As the proportion of the population engaged in farming fell, it is possible that the general level of farming competence rose, as increased competition encouraged the less successful to leave the industry. The change from farming regimes where decisions were communally based, with a strong element of custom and tradition, to regimes based on the individual farmer (discussed in the next chapter) may have given the more able farmers more scope to improve their methods and widened the gap between good and bad farmers.

Important though this issue may be, it is almost impossible to discover farmers' levels of competence. However, we do have the evidence of the extensive body of contemporary literature devoted to agriculture and husbandry, dating from the sixteenth century. This material has been much used by historians as evidence of farming practice, but what was its value for contemporary farmers? There can be no doubt that some authors did offer sensible advice, but on the other hand others made quite bizarre recommendations. The first systematic investigations into agricultural practices were made by the 'Georgical Committee' of the Royal Society appointed in 1664, and the latter half of the seventeenth century saw an upsurge in writing about agriculture. A relative explosion of publication took place from the third quarter of the eighteenth century, initially through the works of Arthur Young, and subsequently through the publications of the Board of Agriculture (of which he was Secretary), founded in 1793. Two *General Views* of the agriculture of each county were published by the Board, and Young edited forty-five volumes of the journal *Annals of Agriculture* from 1784 to 1815. The Board was disbanded in 1822, but by this time many new publications were appearing, including the *Farmers' Magazine* from 1800 and the *Mark Lane Express* from 1832.

Historians continue to debate the merits of this material and in particular the capabilities of Arthur Young. The general consensus seems to be that he judged areas by the criteria with which he was familiar and was therefore blind to other improvements which were appropriate for areas about which he had little knowledge. It is also clear that because his output was so prodigious his observations were not as careful as they might have been. The correspondence of George Culley, the livestock breeder, indicates he thought Young totally incompetent as a farmer. An historian of the agriculture of Lincolnshire concluded, 'It is clear from a careful examination of his work that he often had not visited some of the places he describes so vividly.' In Ireland, according to a contemporary, 'He went with the rapidity of an express, asked for answers to a set of questions and seemed not to notice anything else; seemed, I hear, for I did not see him, very ignorant, not communicative, and to pay equal regard to the assertions of all persons.' Young was not alone in receiving this sort of criticism. In the 1740s the Swedish botanist Pehr Kalm was not impressed with William Ellis, the agricultural author of Little Gaddesden in Hertfordshire, since 'Through this assiduous book-writing it happens that his arable and meadows are worse cared for than his neighbours''. As far as his books were concerned, 'The worst is, that one cannot build upon what is said in them; for he has been too credulous, and has taken as true what false and made-up stories his mischievous neighbours often amused themselves by telling him – of which several persons assured me.'

Despite these verdicts, the work of Young and other eighteenth-century writers does indicate changed attitudes to farming. Young considered 'Experiment is the rational foundation of all useful knowledge: let everything be tried', and indeed he published the results of more than 500 experiments in his *Farmer's Tours* of the 1770s. William Marshall, a contemporary of Young, published as his second book in 1779, *Experiments and observations concerning agriculture and the weather*. Another change is also evident in comparison with the seventeenth-century agrarian literature. Eighteenth-century writers were much more concerned with the profits of farming than their seventeenth-century counterparts whose main interests lay in the activity of farming itself. Young's *Farmer's guide in hiring and stocking farms* of 1770, for example, is devoted to calculating the profits to be gained on farms with differing enterprises and of various sizes. By the mid-nineteenth century farmers had a wide range of farming journals and newspapers available to them. One estimate considers that the minimum weekly readership of the leading titles was 17,000, amounting to half the more substantial tenant farmers and landowners in the country.

This discussion of farming literature as a guide to farming skills is rather inconclusive; in any case its advice was restricted to those farmers able to read. Most farmers would have gained new information from their neighbours, and the generality of farmers did not have much of a reputation for innovativeness. Formal agricultural education was not established until the nineteenth century and the founding of the Agricultural College at Cirencester in 1845, although Samuel Hartlib had proposed the foundation of a 'Colledge of Husbandry' as early as 1651, and in 1790 William Marshall proposed the creation of 'rural seminaries'. Education and information were also provided by local farmers' associations. The earliest foundations were in the 1770s and by 1803 Young listed twenty-three of them. Nearly one hundred were listed in 1835 and over 700 by 1855. These institutions created small libraries, hosted talks by experts, and organised shows. Other agricultural societies were grander in scale and dominated by aristocratic patronage. The most famous is the Royal Agricultural Society of England, founded as the English Agricultural Society in 1838, but two of the earliest were the Bath and West founded in 1777 and the Royal Lancashire founded in 1767. By 1850 the need for agricultural education was clearly recognised: as Pusey put it: 'If our farmers will inquire what is done by the foremost of them, they will themselves write such a book of agricultural improvement as never was written elsewhere, in legible characters, with good straight furrows, on the broad page of England.'

Conclusion

Despite the different ways in which they were constructed the three estimates of agricultural output employed in this chapter all suggest that output increased by some two and a half to three times between 1700 and 1850 (Table 3.11). A greater proportion of that increase was brought about by increased land productivity (Q/T) than by extensions to the area of farmland, although the sown arable area doubled. Extending these figures back before 1700 is very difficult, and is only possible for population-based measures of output, and the partial measure of land productivity, crop yields per acre. It nevertheless appears that annual growth rates of output in the sixteenth century were high, as was the growth in wheat yields, although that growth was not sustained.

Judging by the evidence of labour productivity, which probably fell from the mid-sixteenth to the mid-seventeenth century, the sixteenth-century land productivity rise was probably due to an increase in labour inputs to agriculture coupled with increased reclamation of land for arable farming. Studies of medieval agriculture using demesne accounts, which include information on labour inputs (and are unavailable for the early modern period), have shown that population pressure in the early fourteenth century encouraged farmers to increase output per acre by using more labour. But the price for this increased output was a reduction in labour productivity, and the same phenomenon occurred in the late sixteenth century. Overall, it is unlikely that total factor productivity rose during this period. From 1650, however, the evidence shows that both land and labour productivity were rising together. Between 1700 and 1850 labour productivity in agriculture doubled.

The reasons for increases in output and land productivity are considerably more complicated than the simple model outlined in Chapter One. Land reclamation, changes in the ratio between pasture and arable, a reduction in fallows, the cultivation of fodder crops, changes in the mix of food crops, and regional specialisation were all responsible for raising output, and, to varying degrees, in raising overall land productivity. It is impossible to weigh up the relative contributions of these factors to increased output since the evidence to do so is simply unavailable. Even if it were, most factors were interlinked and their individual effects would be impossible to disentangle.

We are on more certain ground with increases in output per sown acre, which can clearly be related to the management of nitrogen, and especially to the addition of new nitrogen from legumes. Increases in livestock productivity were related to improvements in fodder output per acre, mainly as a result of replacing low-yielding pasture with improved grass, and the extension of arable farming including high-yielding fodder crops. The quality of animals improved, and it is probable that the turnover of animals also increased.

When these changes are considered together, in terms of farming systems, the key development was the integration of grass and grain and the ability to support a higher density of livestock while simultaneously extending the arable area. The integration of grass and grain, at first through convertible husbandry, and later through rotations based on the principle of the Norfolk four-course rotation, made more efficient use of nitrogen for arable crops, and expanded the output of fodder per acre. In Norfolk, for example, the proportion of the arable area sown with fodder crops was between 13 and 17 per cent for the entire period from *c.*1250 to *c.*1730; by the 1830s it was over 50 per cent. The integration of grass and grain probably began in the sixteenth century, and took the form of ley husbandry in the seventeenth, but it was not until the late eighteenth, with the intensive cultivation of turnips and clover, allied to the expansion of the sown arable acreage, that significant and unprecedented growth in output and productivity occurred. The implications of these changes for ideas about the 'agricultural revolution' will be considered in Chapter Five.

In comparison with the material available for discussing changes in output and land productivity, the evidence currently available for analysing labour productivity is meagre. There is disagreement as to the course of labour productivity from the sixteenth century, and the attempt to explain developments in labour productivity amount to little more than a research agenda. However, it is clear that labour productivity was rising from the mid-seventeenth century, that it was rising in conjunction with land productivity, and that it continued to rise under pressure of population growth in the eighteenth and early nineteenth centuries, whereas it had failed to do so with the population growth of the sixteenth century. It is also clear that labour productivity increased without conspicuous technological innovation – new tools and machinery were a nineteenth-century phenomenon – so that much of the explanation for the rise in labour productivity must lie with institutional factors. These are the subject of the following chapter.

4

Institutional change, 1500–1850

Changes in the institutional context under which farming was carried out are important for their impact on output and productivity, but they are also of great significance in their own right because they are concerned with the structure and organisation of agricultural production and with the lives of those working on the land. The bulk of this chapter is taken up with land-holding and social relationships in the countryside. This involves considering changes in the ways in which land was held; particularly the establishment of leasehold as the almost universal form of tenure by the nineteenth century, and the elimination of most common rights. These two processes were often, but not necessarily, associated with the process of enclosure. Enclosure also influenced the process of social differentiation; that is changes in the relative size and importance of different social groups, particularly in their relation to the holding of land. In a nutshell, by 1850 most of the land of England was farmed by tenant farmers under conditions of private property, renting their farms for a period of years from landlords, and employing landless labourers to work on their farms. But before considering how these conditions evolved, the chapter begins with a survey of the main developments in the marketing of agricultural produce between the sixteenth century and the nineteenth century.

The market

A market is a place that provides a forum for the meeting of buyers and sellers who exchange commodities for money. Markets thus serve to gather products from scattered sources and channel them to scattered outlets: that is they redistribute commodities from producers to consumers. In the early modern period markets were held regularly and frequently (once or twice a week) and were distinguished from fairs which were held much more infrequently (once or twice a year), drew both buyers and sellers from much larger areas, and usually handled a much wider range of goods. The area

133

from which buyers and sellers came to a market is also referred to as a market; although more specifically it is the market area or the hinterland of a market. As well as referring to a place or to an area the word 'market' can also be applied to a particular commodity (as in 'the market in wheat') meaning that the commodity has a price and is exchangeable for money.

Markets in these three senses were commonplace in 1500. There may have been as many as 800 market towns in England, redistributing food supplies from the farms in their hinterlands to consumers, although it is impossible to say how active each of these markets was. Most agricultural commodities also had a market: they were tradeable and had a price. Yet despite having these modern characteristics, markets and marketing in the early sixteenth century were fundamentally different from the situation in the mid-nineteenth century. In the early sixteenth century most markets (as locations) were strictly controlled by rules and regulations governing the behaviour of buyers and sellers. Thus, although markets were 'open', they were not 'free' in that prices were not determined by the free interplay of supply and demand. Regulation and custom also played an important part in determining price levels, so although virtually all commodities had a price, those prices were not determined solely by the competing bids of buyers and sellers. By the nineteenth century, markets were not so regulated and were therefore 'free': prices were 'bid-prices' determined by the competitive bidding of buyers and sellers. Although it was often technically illegal, much marketing activity took place away from the formal area of the market place, and deals were struck between individuals in private. These private deals were not usually between the direct producers (farmers) and consumers, but between producers and middlemen or wholesalers in grain. These middlemen were the object of much regulation but they came to dominate the process of agricultural marketing by the nineteenth century. In terms of its operation the most important developments in marketing between the sixteenth and the nineteenth centuries were therefore the rise of the middlemen and the development of private marketing.

These developments were reflections of a dramatic increase in the scale of marketing, both in terms of the proportion of produce traded and in the distances over which it was traded. By the nineteenth century, the vast majority of agricultural production was for the market – for exchange rather than for use by the farmer and his family. By this time the market in most agricultural commodities was nation-wide: local market networks were linked together so that we can speak of a national market in most agricultural commodities. This implies considerable improvements in the movement of commodities and in the flow of information about the conditions of their supply and demand. Before considering the widening of the market, however, we need to examine the situation at the start of the sixteenth century.

The regulated open market

In the early sixteenth century the market was looked upon as a place where local food surpluses could be distributed to those who needed them. The contemporary notion of the 'ideal' market was one in which farmers brought all their surplus food for sale in the open market, and all exchanges took place in the market directly between farmers and consumers. Middlemen, who interspersed themselves between producer and consumer, obviously disturbed this ideal and were generally reviled. By intercepting food from the farmers and preventing it from being available to those who needed it, middlemen were thought to keep prices high and to be responsible for dearth. One form of market regulation therefore attempted to control their activities. Regulation was enacted at both the national and local levels. Parliamentary acts of 1552 and 1563 stipulated that all middlemen should be licensed, and as late as 1650 these laws were tightened by further legislation. Other examples of national legislation include the Assize of Bread, which attempted to regulate the weight and price of bread. This dates back to the middle ages and there is abundant evidence throughout the early modern period of attempts to enforce it at a local level. Other legislative attempts at controlling trade were prompted by shortage. The 'Book of Orders for the Relief of Dearth' was issued by the Privy Council to counteract the dearth of 1586–7, and implemented periodically thereafter. Local commissioners were empowered to force those hoarding supplies of grain to bring them to the market, and to regulate the activities of the middlemen.

At the local level each market town had its own regulations and its company of market officers to enforce them. Their activities involved the testing of measures, balances and weights; the control and licensing of middlemen and other traders; the fixing of prices of bread, malt, meal and corn; the prevention of civil disobedience; and the levying and collecting of market tolls. These tolls provided the funding for these market officers, but the surplus produced could be used for charitable purposes or simply entered the coffers of the town government. The exact nature of regulation varied from place to place, but some idea of the impact of local controls can be gained, through those applying to a single market in Wiltshire as they were set out in March 1564 (the language has been modernised).

1. Before the market starts the sellers of grain are to agree with the local justices what the price should be.
2. No transactions may take place before 9 a.m. when a bell will be tolled 20 times.
3. When the market opens purchases must be for the customer's own use and be limited to 2 bushels of grain.
4. After 11 a.m. (when the bell is again tolled 20 times) grain may be bought by those who will resell it (eg. bakers, brewers and badgers).

5. Those buying grain to resell must be licensed by a Justice of the Peace.
6. Grain may only be bought on market day.
7. No person may buy grain in the market if she has sufficient of her own.

These rules illustrate the restriction of the activities of middlemen and the attempt to force all dealings into the formal market. But they also show that not all consumers were assumed to have equal weight in the market place, since purchases for the first two hours of the market were to be made by those buying relatively small quantities of grain. We do not know how strictly these regulations were enforced, whether the formal structure of the market matched the economic reality, but we do know that they are typical of those found in markets in many other parts of the country. Moreover we can find evidence of their application as late as the late eighteenth century. However, it seems that they tended to be enforced only during periods of dearth when grain was in short supply and prices were high. In normal years the protection they offered was not needed. Regulation of open markets provided protection for consumers, especially in conditions of dearth, but it also helped protect farmer and consumer alike from fraud and bad practice. The education of most early sixteenth-century farmers was inadequate to protect them from fraud. They were illiterate, and even if they were reasonably numerate, the recording of quantities and prices in roman numerals (which was common until the early seventeenth century) offered plenty of opportunity for fraud and sharp practice.

Yet the impact of regulation must not be exaggerated. Formal market regulations and their administration are conspicuous to historians because they were documented. It is very likely that a considerable amount of marketing took place outside formal, regulated, markets. A study of the trading network of Warwickshire and Worcestershire in the middle ages has shown how much trading was conducted outside the formal framework of boroughs and markets. This 'hidden trade' was no doubt still in existence in 1500, and consisted of trading in towns and villages not subject to control by a lord, at country fairs, and other places where people congregated such as inns located on the road network, and at ports.

Market development, 1500–1850

The gradual breakdown of regulation and the other changes in the character of marketing were prompted by the dramatic growth in the volume of market activity from the sixteenth century. This cannot be measured directly but it is implied by rates of population growth, urbanisation (especially the growth of London), and the growth in the proportion of non-agricultural workers in the countryside. Increased market activity was the corollary of the decline of subsistence farming and also the consequence (and to some extent the cause) of regional specialisation in agricultural production. The widening of the market is also evident in changes to the marketing process: the rise of the middleman and private marketing.

Most market activity at the start of the sixteenth century, whether carried out within formal or informal networks, was 'local marketing', with markets having a hinterland with a radius of about 10 miles. It has been estimated that such exchange systems could support towns of up to about 10,000 people. These local systems were 'closed' in the sense that relatively little of the produce was traded outside the local market area. The next stage in market development took place when trade developed between these local market areas. Such inter-market trade is significant because it means that middlemen became an essential part of the marketing process, buying in one market and selling in another. It also follows that middlemen dealing between markets needed good information about supply, demand and prices in distant markets. At the same time the middlemen trading between markets often did so without the grain being physically present, which in turn led to such developments as the forward buying of grain. Transactions of this nature were facilitated by the development of systems of long-distance credit which required different institutions from the ones sufficient to support local systems of borrowing and lending.

The third stage in market development came when a more fully integrated national system of exchange arose, rather than simply trade between individual markets. Such a national system, where, say, London corn merchants travelled the country in search of grain to be purchased in advance of the harvest and shipped directly to London was certainly in place for the major agricultural commodities by 1800. The movement from one stage to another in this sequence of market development, from predominantly local marketing to inter-market trading, and then to national trading, was prompted by an increase in the volume of produce traded, which was probably caused by an increase in demand from outside a local market area. In the early modern period, but especially in the seventeenth century, much of this demand came from London.

English population growth is shown in Figure 3.1 and Table 3.5(a). All else being equal, the rate of market activity must have risen at this rate. However, the growth of market activity was proportionately greater than population growth because a growing proportion of the population was living in towns (necessitating rural to urban transactions) and an increasing proportion of those living in the countryside were not producing food (necessitating rural to rural transactions). The proportions of the population living in towns (of over 5,000 inhabitants) and living in the countryside but not engaged in food production are shown in Table 4.1. This indicates that the proportion of people not working in agriculture (and therefore dependent on the market for their food) at least doubled between the early sixteenth century and the early eighteenth century.

Table 4.1 *The English non-agricultural population, 1520–1851 (percentages)*

	Towns > 5000	Rural non- agricultural	Total	London
1520	5.25	18.50	23.7	2.25
1600	8.25	22.00	30.2	5.00
1670	13.50	26.00	39.5	9.50
1700	17.00	28.00	45.0	11.50
1750	21.00	33.00	54.0	11.50
1801	27.50	36.25	63.7	11.00
1851	48.00	30.60	78.6	14.00

Of more importance than the general trend in urbanisation was the growth of London. In the early sixteenth century London had a population of around 55,000; this grew to 200,000 by 1600, 575,000 by 1700 and 960,000 by 1801. The proportion of the population living in London is shown in Table 4.1, but the impact of London on the demand for food was greater than these figures indicate because average consumption per head in London was at least double the national average. Finally, the growth in regional specialisation must also have contributed to the growth in market activity. This process has been described in the previous chapter, where it is pointed out that it is difficult to measure except in a general way using indirect evidence. As specialisation developed, increased market activity was initially located at the boundaries of specialising regions.

Even in the early sixteenth century, before the onset of rapid population growth and urbanisation, some agricultural produce was moved over long distances. Indeed in 1300 the population of London had been around 80,000, and food came by road from up to 20 miles away and by water from up to 60 miles away. The city took some 10–15 per cent of all the food produced for human consumption within this hinterland. In the sixteenth century complicated marketing arrangements involving agents (known as factors) buying food on commission existed for the provisioning of the army and navy, and for the Royal Household. Every year, agents for the Royal Household of Henry VIII had to purchase something in the order of 1,500 cattle, 8,000 sheep and 3,000 quarters of wheat. There were farmers in the sixteenth century whose scales of enterprise were too great to be handled by the conventions of the local market place. The surviving farm accounts of Peter Temple of Burton Dassett in Warwickshire show that in the 1540s he bought cattle to fatten from Wales through markets at Chester and Shrewsbury. For the years 1545–50 the accounts show that although some beasts were sold locally, the majority were sold at the farm (rather than at the market) to butchers directly supplying the London market. Sheep too were sold at the farm, sometimes in large quantities, for the London butchers, but also to other large sheep farmers such as the Spencers of Wormleighton in Warwickshire.

The initial widening of the market was more likely to occur with livestock and livestock products than with crops. Live animals could be walked to market, and butter, cheese and wool were high in value relative to their bulk and hence better able than grain to withstand the costs of overland carriage. This meant that livestock were less dependent than crops upon proximity to markets or cheap water communications for their commercial development, and ensured that pastoral production in relatively peripheral regions became increasingly geared towards the demands of the core. By the seventeenth century, livestock marketing had become organised on a national scale. Welsh cattle had been driven to England since the sixteenth century, where they were dispersed to fattening regions all over the south of England (see Figure 2.5). From the mid-seventeenth century they were joined by those from Scotland, mainly from Galloway, but including some from the highlands. Considerable numbers of cattle were also imported from Ireland until the trade was banned from 1667. By the late eighteenth century this trade was extensive and well organised. For example, a contemporary reckoned that one half of Norfolk cattle were driven from Scotland, a quarter from Wales and Ireland, with only the remaining quarter being bred in the county. The poet John Clare described a cattle drove as follows:

> Along the roads in passing crowds
> Followd by dust like smoaking clouds
> Scotch droves of beast a little breed
> In swelterd weary mood proceed
> A patient race from scottish hills
> To fatten by our pasture rills
> Lean wi the wants of mountain soil
> But short and stout for travels toil

Traffic in livestock intensified as urban populations grew during the eighteenth century. By the eve of the railway age, a truly national market in livestock existed, with further developments in regional specialisation of the various stages of livestock production.

These developments in the livestock trade illustrate inter-market trade, that is exchange between local markets, and its progression towards national marketing. Similar developments took place in the marketing of grain, but at a slower rate. In the sixteenth century, grain was moved between markets, but there was no national market and, with the exception of the provisioning of London, most grain movements were within local market areas. In the early seventeenth century, Henry Best, the east

Yorkshire farmer, was clearly farming to make money and knew exactly which markets would give him most profit. Yet his grain was sold locally to a number of markets in quite small quantities at a time. He was aware that, when information and capital were available, large-scale trading in grain could be profitable although it does not seem to have been the norm. Thus he tells a story about a York man who bought 3,000 quarters of barley in Norfolk at 13s a quarter and shipped it to York, when at Driffield, Best's local market, barley was being sold for 20s per quarter. If we take this anecdote at face value it illustrates that wide price variations could exist in the early sixteenth century, that the exploitation of these was unusual, but that Best was alive to the possibilities that such trading might have had for making money.

Although many farmers were still selling to local markets in subsequent centuries, the quantities of grain moved over longer distances expanded enormously. By 1700, over one million quarters of grain (getting on for 200,000 tons) was being consumed in London for food and drink, and the regular hinterland of the capital for grain now included coastal Sussex, Norfolk and even south Yorkshire. By the early nineteenth century its supply was truly international, with grain being supplied from Friesland, Holland and even southern Sweden. While undoubtedly the most important urban influence upon the marketing system was London, other towns also provided a strong demand, especially after the 1780s as the towns of northern England began to grow extremely rapidly.

On the basis of the evidence of regional price movements it has been suggested that there was a national market in wheat by 1700. By the 1830s, the market in most agricultural commodities, excluding only those perishable products like liquid milk and vegetables, was a national one; and indeed was international in that it stretched across much of northern Europe. It is possible to demonstrate market integration through the statistical analysis of regional prices, but it is also evidenced in other ways. In his study of English agriculture in the mid-nineteenth century, James Caird compared the prices of bread and meat by county with those given by Young in his *Tours* published some seventy to eighty years earlier. Caird's bread price was constant across all counties, at a penny farthing, while Young's varied from three farthings to 2d. The same pattern was also true for meat, with Caird's universal valuation at 5d per pound and Young's varying between 2½d and 4d. Market integration in the early nineteenth century is also evidenced by the increased convergence of regional differences in weights and measures.

The widening of the market signalled the end of local subsistence crises. Such subsistence crises were not frequent during the sixteenth century but they were not unknown. Periods of acute shortage following bad harvests were in the summer and autumn of 1557, from the late summer of 1558 to the winter of 1559, from 1596 to 1598, and during 1622–3. In the absence of direct information on the causes of death during these crises it is difficult to be certain that they can be attributed to famine, but it seems very likely that in Cumbria, for example, there were mortality peaks caused by famine in 1587–8, 1597 and 1623. Price and other evidence suggests that the most severe crisis of the sixteenth century came over the winter of 1596–7, yet only 18 per cent of parishes in the Cambridge Group's sample of 404 English parishes experienced a crisis of mortality. Crisis mortality was confined to certain areas: those in remote highland areas, or in lowland areas of predominantly non-arable farming with large populations of poor people. The last period for which we have clear evidence of famine and a subsistence crisis was 1622–3, when people starved to death in Cumberland, Westmorland and perhaps also some Durham parishes. Yet paradoxically these remote areas were not subject to famine because they were outside a national marketing framework: their small farmers were the victims of premature specialisation. An economy based on the production for the market of livestock and animal products and the purchase of grain for food from the market broke down when extreme price fluctuations in periods of dearth meant that grain prices were much higher than livestock product prices. As a consequence, small farmers had insufficient income to buy food. On occasion short-term difficulties of transport could exacerbate a shortage of grain even in areas that were normally accessible. A mortality peak in the Durham parish of Wickham (just across the Tyne from the major city of Newcastle) over the winter of 1596–7 seems to have been a crisis of subsistence. Mortality was concentrated in the winter months of November to March when local food supplies were inadequate and external supplies could not reach the north-east ports: by July 1597 rye was being sold for 96 shillings a quarter when its average price for the 1590s was just over 10 shillings. It is difficult to distinguish here between famine as a consequence of poverty, famine as a consequence of an inadequate national production of food, and famine as the consequence of an inadequate marketing system for distributing food within the country. One way in which this can be investigated is through regional price trends, but, as yet, we have insufficient evidence for the sixteenth century.

The growth in the volume of marketed produce went hand in hand with the development of the transport network, but such developments were also important in improving the flow of market information, that is information on supplies, demand and prices elsewhere. The cheapest way to move grain was by water, either using coastal shipping or by inland waterway. Between 1600 and 1702 there was a two-thirds increase in the tonnage of coastal shipping, but port books indicate that the volume of corn shipped to London via the coast *c.* 1700 was only about 20 per cent of the total carried to the capital. Trade by road, but more especially by river, was much more important. The seventeenth century saw considerable improvements in river navigation. There were about 700 miles of navigable rivers in England in 1660, 900 miles by 1700 and some 1,100 miles by the 1720s. Sizeable barges reached Reading on the Thames by the late sixteenth century and were reaching Oxford by the mid-seventeenth century. By 1677 the Severn was navigable as far upstream as Welshpool. The most spectacular river development was the Wey navigation in Surrey from Guildford to Weybridge, improved from 1651 to 1653 with its ten locks.

There were also considerable improvements to the roads, and it has been estimated that between 1500 and 1700, the capacity of English road transport grew three- or fourfold. Large waggons could transport grain overland, but commodities with a higher value to weight ratio such as wool or cheese could be economically transported by teams of packhorses. After 1700, roads were developed by turnpike trusts which improved lengths of road and charged users a toll for using them. There were almost 15,000 miles of turnpikes by 1772 and about 20,000 miles by the mid-nineteenth century. Roads were necessary to the growing national market for distributing small quantities of goods over a wide area. In the middle ages it has been reckoned that the transport of grain by road was about twelve times the cost of water transport: it is testimony to the improvements in roads that this proportion had reduced to five times by 1700.

The cutting of canals started in earnest in the 1770s, so that by 1772 the Mersey was linked to the Severn, the Trent to the Mersey by 1777, the Severn to the Thames by 1789, and the Mersey to the Trent and Thames by 1790. Canals undoubtedly exposed more areas to the possibilities of the long-distance grain trade, but they also had more local impacts. The Bridgewater canal opened the Merseyside market to potatoes grown in Cheshire; and fruit growers in Worcestershire used the Staffordshire canal to send apples to Lancashire. Canals were also important for the trade in farm inputs for they enabled the cheap transport of lime and manure to farms adjacent to them.

The arrival of the railways in the 1830s further reduced transport costs, although it was not until the second half of the nineteenth century, with the proliferation of rural branch lines, that the railway had its major impact on the farming community. The speed of the railway changed marketing practices for a number of commodities and created new opportunities for farmers as a consequence. Perishable products, such as liquid milk and market garden produce, could now be produced at some distance from consumption centres, although again these developments were to have their major impact in the second half of the century. Railways also became important for the movement of livestock, replacing the long-distance droving of cattle and sheep.

As well as speeding the movement of goods, developments in transport improved flows of information. By the late seventeenth century, John Houghton was able to publish weekly grain prices from around the country in his *Collections for the improvement of husbandry and trade* using information brought to London by post horses. Mail was also carried by passenger coaches: 900 services a week left London in 1715 and on the eve of the railway age in 1830 this figure had grown to 1,500 and in 1838 over 60 million letters were sent before the advent of the penny post in 1840.

The widening of the market was accompanied by changes to the marketing process; especially the growth in middlemen and the rise of private marketing; and changes in the regulation of markets which reflect a changing ideology of 'the market'. Middlemen performed a number of functions which were essential for the expansion of the marketing system. The ideal model of a market acting as the means for farmers' surpluses to be redistributed to local consumers breaks down once local supplies are either too great or too small to match local demand, and the means are available for physically removing produce to other markets in the case of oversupply, or of importing food in the case of dearth. These functions would have to be performed by wholesalers and dealers. It was presumably for this reason that the northern counties of Cumberland, Westmorland, Lancashire, Cheshire and Yorkshire were exempt from a 1563 statute against middlemen. These regions were deficient in grain production and needed corn dealers to provision them with supplies from outside. Middlemen were necessary for supplying the largest cities whose hinterlands spread much further than the normal range of a farmer's nearest market. They were also necessary to link local markets together to supply cities such as Bristol, York, Norwich and Exeter, and especially London. From the supply side we have seen that in the sixteenth century there were already farmers selling large amounts of grain or large numbers of livestock. Local markets were simply inadequate for these farmers because they had insufficient buyers, so they had to sell their produce to middlemen.

Despite hostility from most quarters, middlemen gradually became indispensable to the grain trade, so that by the nineteenth century it was the exception rather than the rule for producers of food to sell it directly to consumers. There were several categories of middleman distinguished according to the function they performed. In grain markets the 'forestaller' bought corn before it came to the market, the 'regrator' bought corn in the market not for his own use but in order to resell, and the 'engrosser' withheld corn from the market and stored it until prices rose. Middlemen were referred to with a variety of names such as 'badgers', 'kidders' and 'jobbers'. For the marketing of livestock the main middleman was the drover, who bought and sold cattle and drove them to market.

As the seventeenth century progressed, the functions of some middlemen became more specialised, but at the same time those whose primary function was in processing food, including millers, bakers, maltsters and butchers, became more closely involved with dealing in grain or livestock. Middlemen also became increasingly important for their financial activities, especially in giving credit in the seventeenth and early eighteenth centuries, when banking was rudimentary. By the nineteenth century middlemen were a well-established and integral part of the marketing system. The census of 1851 records about 250,000 'farmers', and over 25,000 people employed specifically as agricultural middlemen (including corn merchants, cheeseburgers, cattle dealers and pig dealers), but this understates the number of middlemen who were classed by the census as primarily maltsters (11,150), and as millers (37,268).

The regulations of the open market, and their tolls, discouraged dealing by farmers and middlemen in formal markets. Thus long before the grip of regulation began to lapse, goods were increasingly exchanged outside the open market – a phenomenon known as private marketing. Some of these transactions might have been between producer and consumer, but the majority would have been between the farmer and middleman. Increasingly farmers took a sample of their grain in a small bag to show the corn dealers and sale by sample gradually became the most common way for farmers to sell grain; isolated examples are evident in the seventeenth century or even earlier, but the practice was common by 1750, especially in market towns near London, and the practice was fully developed by the 1830s. In London, corn dealers erected a corn exchange in Mark Lane in 1750 where selling by sample occurred. Defoe described the corn marketing process as follows: 'Instead of the vast number of horses and wagons of corn on market days there were crowds of farmers, with their samples, and buyers such as mealmen, millers, corn-buyers, brewers etc., thronging the market; and on the days between the markets the farmers carried their corn to the hoys and received their pay.'

By the turn of the nineteenth century the inn had replaced the market place as the locus of corn dealing in most towns, and by mid-century many towns had built elaborate corn exchanges as the locus for grain sales by sample. Selling by sample encouraged the buying of large quantities of grain, and frequently corn dealers would contract to purchase the entire grain crop of a farm. This could be after the harvest following a sale by sample, but forward buying also developed as corn dealers toured farms and agreed to purchase an entire crop before it was harvested. Such practices were common in East Anglia in the early eighteenth century when corn merchants purchased the barley crop on the farm for shipment to London for malting or to the East Anglia ports for export to the continent.

As private marketing developed, the number of open markets seems to have declined and by the mid-eighteenth century the number of towns holding a market had fallen to around 600. Although some markets had been specialising in certain commodities in the sixteenth century, by the eighteenth century levels of specialisation had increased considerably. The decline of fairs was even more pronounced, but some continued to prosper well into the nineteenth century, especially for livestock. Changes in the marketing process also put pressure on the more rigid forms of market regulation. Regulation was also less necessary as the likelihood of dearth diminished from the mid-seventeenth century and as the education of farmers improved. Business methods were still rudimentary and farm accounts were the exception rather than the rule, but by the mid-eighteenth century farmers and consumers were in less need of protection against unscrupulous dealers. The expansion of market areas also reduced the likelihood that local suppliers could create a monopoly which would be acting against the consumers' interests.

The turning point in government attitudes came in the 1660s: whereas in the sixteenth century there were continued attempts by the state to regulate marketing, after 1663 people were permitted to regrate and engross (but still not to forestall). By the eighteenth century the enforcement of existing legislation against middlemen had become weaker. In Manchester, for example, during 1650–87 there were about 185 market offences per decade brought to the court leet, but by 1750 only 23, and they were mostly concerned with adulterated food. A resolution of the House of Commons condemned legislation against forestallers, regrators and engrossers and existing legislation was repealed in 1772. Finally, the Assize of Bread was abolished in London in 1822 and for the rest of England in 1836. Nevertheless there were repeated attempts in the eighteenth century to enforce legislation for market regulation. An attempt was made to revitalise the Assize of Bread, for example, in 1710, which was modified in 1758, and again in 1772. Certainly the poor were vociferous in their desire for traditional regulation in times of dearth.

On the other hand, by the late eighteenth century the idea of the 'self-regulating' market gained currency, so that opposition to regulation acquired some intellectual justification. This is important because it marks a fundamental shift in attitude towards the nature of the market. Thus high prices following a bad harvest were seen as a 'natural' form of rationing and consequently no intervention was needed. This new ideology ignored the fact that the rationing mechanism was not equitable since the burden of rationing fell disproportionately on the poor. This contrasts with the interventionist tradition which had attempted to protect the weakest consumers from the worst consequences of dearth.

While the government repealed legislation controlling domestic marketing by the end of the eighteenth century, legislation affecting foreign trade continued into the nineteenth. Statutes controlling the movement of grain overseas were enacted in the fourteenth century, and in the sixteenth century the principle was established that exports would be permitted when the domestic price fell below a stipulated level. From the 1660s exports were subsidised and imports limited as part of a policy which is known as 'mercantilism'. Starting in 1672, a series of acts known as the corn laws subsidised exports provided the domestic price was below a certain level. In practice, prices did not often fall below this threshold, but the policy remained in force until 1846. The export of malt also received a subsidy. Between 1697 and 1760 those exporting malt were entitled to claim back the excise duty paid on malt for home consumption of 4s per quarter, and therefore received a subsidy irrespective of the domestic price. A new phase in the history of the corn laws began in 1815 in that policy was now much more specifically directed at helping the producer rather than the consumer of grain. Thus exports were permitted whatever the domestic price, but foreign wheat could not be sold in English markets until the price rose above 80 shillings a quarter. By this time the corn laws, as part of the wider issue of commercial protection through tariffs, were a highly contentious political issue. As prices slumped after the Napoleonic Wars the issue became dormant again, but in the late 1830s a rise in food prices and an industrial depression brought the question to the fore once again, and in 1838 the Anti-Corn Law League was formed with the single objective of a free trade in corn. Eventually, in 1846 the government introduced a bill to repeal the corn laws, so permitting free trade.

In retrospect the corn laws are important for their role in short-term political machinations, and in the longer term as an indicator of the changing balance of political power between the industrial and agricultural interests in the country. The consequences of repeal, at least before the 1870s, were not as great as the landed interest had feared. Prices remained at former levels and world grain prices rose rather than English grain prices falling. Livestock prices did, however, rise by 30 to 40 per cent relative to grain prices, which encouraged a move away from grain production, although on a scale that was slight compared with the reaction to the more extreme price differentials in the last quarter of the century.

By the early nineteenth century there can be no doubt that marketing functioned in a radically different way than it had in the sixteenth century. Moreover the ideology of markets and marketing had changed so that 'free' markets were now acceptable whereas they had once been illegal. As a consequence, producers were now separated from consumers, so whereas in the sixteenth century farmers and consumers met in the market place, by the nineteenth century both dealt with middlemen: one with the corn merchant and the other with the baker. These market developments are obviously related to the changes in agricultural production discussed in the last chapter, but the nature of the relationship is difficult to specify. There is a reciprocal relationship between many agrarian developments and the development of the market so it is often impossible to point the causal arrow in a particular direction. Increasing farm sizes, rising labour productivity, the growth of agricultural wage labour and the increase in regional specialisation could be both a cause and a consequence of market development. It is clear that each needed the other: farmers required the stimulus of increased demand mediated by the market to increase production and profit, but that demand could only continue provided they were able to increase production. Markets became more efficient in mediating between supply and demand as regulation declined and self-regulating markets lowered transaction costs. This widening of the market facilitated an increase in local and regional specialisation within agricultural production, where markets channelled produce between rural areas, and contributed to the process of urbanisation, where markets channelled produce from the countryside to the towns.

Tenures, property rights and enclosure

We have seen in Chapter Two that in the early sixteenth century land was held under a great variety of both tenures and estates. By the nineteenth century these had been transformed so that most land not in the hands of owner occupiers was held under leasehold tenure for a period of years. The three centuries from *c.* 1550 also saw the establishment of private property rights on almost all the arable land of England. Common property rights lingered on pasture land in some parts of the country, but in the main, by 1850, English farming was carried out under conditions of private property, with farmers making their decisions about how farming was carried out individually and without reference to others.

The process by which property rights were changed is called 'enclosure', a blanket term used both by contemporaries and modern historians to cover a variety of changes in landholding. In attempting to understand the causes, mechanisms and effects of enclosure it is important to realise that enclosure itself could involve several processes: the establishment of leasehold, the removal of common property rights, changes in farm layouts and field boundaries, the amalgamation or engrossing of farms, and a radical change in land use. Although these processes were related they were not necessarily part of a single unified process. Some 'enclosures' may have involved little physical alteration to boundaries, for example, and have been primarily concerned with changing rights to the use of property; other enclosures might have been more concerned with a major change in land use (say from permanent arable to permanent pasture). It is also the case that changes in tenures and cultivation practices, and engrossing, could take place without any enclosure at all, indeed some parts of England saw no enclosure in the early modern period, since about 45 per cent of the country was already enclosed in 1500.

Enclosure can also be categorised according to the kinds of lands it affected. An important distinction is between enclosures of commonfield arable and common waste. The former usually removed common rights, might have involved changes to the nature of leases, and when carried out on a large scale could remodel the layout of a village with new fields, new farms and new roads. Enclosure of upland wastes could have involved the construction of new boundary walls and the removal or drastic alteration to common property rights, but little change in land use need have taken place. Enclosure of lowland wastes, such as heathland and marshland, could, on the other hand, result in a dramatic change to more intensive land use, often involving a switch from extensive pasture to intensive arable. Yet another way in which we can classify enclosure is whether it dealt with a whole village at once, or whether it was gradual or 'piecemeal'.

Economic historians writing in the first half of the present century concentrated on two periods of enclosure in England: Tudor enclosures (covering the late fifteenth and the sixteenth centuries) and parliamentary enclosures of the eighteenth and nineteenth centuries. These two periods of enclosure are conspicuous because they attracted the attention of the state. In the first period they were the subject of government commissions because they were held to be responsible for depopulating villages and therefore against the public interest. It is a measure of the change in attitude of the state that parliamentary enclosure, in complete contrast, was enclosure sanctioned by an act of parliament. But in both periods (although more especially the latter) their legacy is a large body of documentation accessible to historians.

Recent research has revised this implied chronology of enclosure in England, so that now the seventeenth century is regarded as having had the fastest rate of enclosure, when about 24 per cent of the country is estimated to have been enclosed compared with only 2 per cent in the sixteenth century, 13 per cent in the eighteenth century and 11 per cent in the nineteenth century. These national figures are in some dispute, but it is difficult to improve on their accuracy. Enclosure by act of parliament is well documented, but earlier enclosures are not. The most detailed evidence is provided by successive bodies of commissioners from 1517 to 1607 enquiring into enclosure and depopulation, but they only provide a very partial record of enclosure and need to be supplemented by evidence from many other sources. Regional variations in enclosure were considerable. Table 4.2 shows the progress of pre-parliamentary enclosure in Leicestershire, where 'Tudor enclosures' are in evidence, although there was proportionately more enclosure in the first half of the seventeenth century. It seems that 25 per cent of the county was enclosed by 1607 and 47 per cent

Table 4.2 *Places wholly enclosed in Leicestershire before 1750*

Date	No.	%
Before 1550	52	36
1550–1600	7	5
1600–1650	57	40
1650–1700	24	17
1700–1750	4	3

Table 4.3 *Enclosure in County Durham, 1550–1850 (percentages of total acreage enclosed)*

Date	%
1551–1600	2
1601–1650	18
1651–1700	18
1701–1750	3
1751–1800	35
1801–1850	24
Total acreage	184,733

by 1710, although enclosure in the period before 1607 was concentrated in two periods: 1485–1530 and 1580–1607. The figures for County Durham in Table 4.3 cover a longer time span, including parliamentary enclosure, and indicate a bout of seventeenth-century enclosures (mainly of open-field arable) and considerable parliamentary enclosure (mainly of upland pastures). Enclosure in the South Midlands (Table 4.4) shows a somewhat different chronology, since parliamentary enclosure was extremely heavy. Nevertheless the period of next most rapid change was indeed the seventeenth century. The exact chronology of enclosure before the eighteenth century must therefore remain in some doubt, but it is evident that there were distinctive regional differences. In any case estimates of the rate of enclosure for the most part fail to distinguish between the different activities that enclosure involved. Documents accompanying the acts for parliamentary enclosure usually show details of the remodelling of the fields and field boundaries, changes in property rights, and whether the enclosed land was arable or waste. Altogether some 4.5 million acres of open-field arable and 2.3 million acres of waste were enclosed by act of parliament; just over 20 per cent of the area of England (Table 4.5 and Figure 4.1).

The general geography of enclosure is more easily reconstructed than its

Table 4.4 *Enclosure in the South Midlands (percentages of total acreage enclosed)*

Date	%
pre 1450	4
1450–1524	6
1525–1574	2
1575–1674	17
1675–1749	5
1750–1849	55
post 1850	3
undated	8
Total acreage	2,850,866

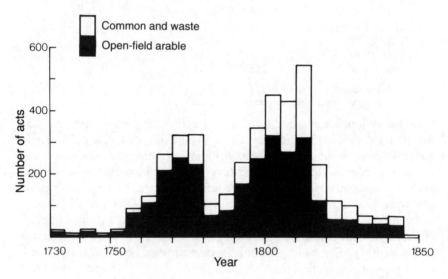

Figure 4.1 The chronology of parliamentary enclosure, 1730–1850 (number of acts per half decade). *Source:* Turner (1980), 68.

chronology. The two maps in Figure 4.2 are from the work of Gonner, first published in 1912, and not yet superseded on a national scale by subsequent research. His maps are based on travellers' descriptions: Leland's Itinerary made between 1535 and 1543 for Figure 4.2(a) and Ogilby's road atlas of 1675 for Figure 4.2(b). Figure 4.3 shows that the core of parliamentary enclosure of arable land (mostly common open-fields) lay in a band

Table 4.5 *The chronology of English parliamentary enclosure*

	Open-field arable		Common and waste	
	Acts	%	Acts	%
1730–9	27	1	12	1
1740–9	28	1	11	1
1750–9	87	3	30	2
1760–9	316	12	77	6
1770–9	481	18	159	12
1780–9	152	6	85	6
1790–9	413	15	163	12
1800–9	591	22	289	22
1810–9	430	16	349	26
1820–9	107	4	109	8
1830–9	79	3	46	3
Total	2711		1330	

stretching from east Yorkshire to Wiltshire, whereas the enclosure of waste was concentrated in the physically marginal areas of the north and west, with pockets of enclosure on the lowland heaths and marshlands.

It is sensible to deal with each of the major changes that could be associated with enclosure in turn: the ending of customary tenures and the establishment of leasehold; the removal of common rights; topographical change and the removal of open-fields; and changes in land use. Following on from this some of the motives for enclosure will be considered and an attempt made to evaluate the impact of enclosure on output and productivity.

The variety of ways in which land was held in the early sixteenth century has been discussed in Chapter Two: most land was held in the form of customary tenure, of which copyhold was the most usual; estates could be for years, lives, or at the will of the lord; and rents were often paid in two parts, a large entry fine, and a much smaller annual rent which was considerably less than the annual value of the holding. By the middle of the nineteenth century most farmers not owning the land they farmed held their land by a lease for a term of years for a rent that reflected the annual value of the holding. The abolition of customary tenures and the establishment of leaseholds for periods of years was often associated with enclosure but there was no necessary connection between the two: a switch to leaseholds for a period of years at rack rents (rents that represented the full annual value of the holding) could arise for many reasons. Such leaseholds were more attractive to landlords because they enabled rents to be renegotiated at the end of the term of the lease, rent reviews were possible at intervals

152

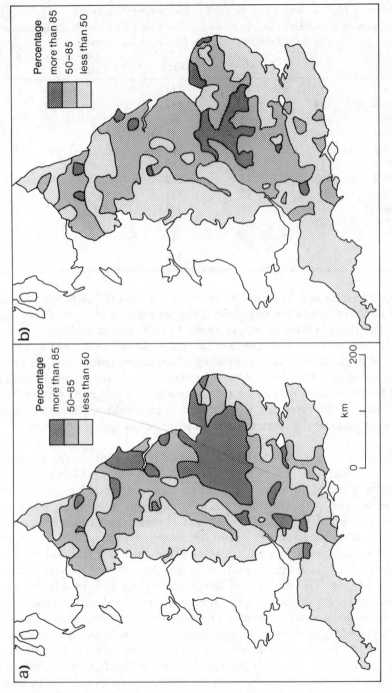

Figure 4.2 Open-field in (a) *c.* 1600 and (b) *c.* 1700. *Source:* Gonner (1912), maps D and C.

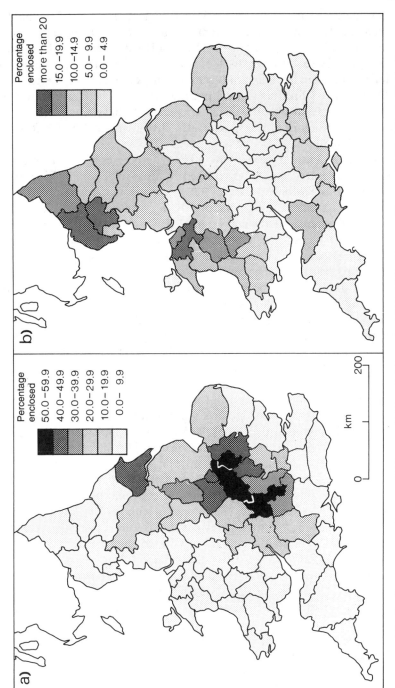

Figure 4.3 Parliamentary enclosure of (a) open-field and (b) waste, 1700–1850. *Sources:* Turner (1980), 180–1; Welsh figures kindly supplied by John Chapman.

during the term, and rack rents reflected the market value of the holding. In contrast, customary tenures often had fixed rents that bore no relation to the value of the holding and could not in law be changed. The incentive to change to leases with shorter terms was especially strong during periods of inflation (in the later sixteenth century for example) since the real value of fixed rents declined as general price levels rose. The converse was also the case, so the conversion to rack rents seems to have slowed down from the 1660s when longer leases were attractive as agricultural prices were stable or falling.

Historians have long been interested in the processes by which customary tenants disappeared and were replaced by leaseholders, and have argued about how this was brought about. One view sees the decline of customary tenants through illegal manipulations of entry fines and forcible evictions by lords who replaced their customary tenants with leaseholders paying rack rents or turned arable commonfields over to sheep walks. Another view argues that customary tenants received the full support of the law against any illegal activities by their lords, and the transition to leasehold is seen as an equitable process owing more to the vagaries of the economics of farming than to landlord coercion. Landlords had a number of opportunities to establish leasehold farms with relatively little controversy. Demesne land was not customary land, and therefore could be leased to tenants for a period of years, and since these were usually new tenancies, they would not involve the displacement of existing tenants. Lords also had the opportunity to convert customary land to leasehold when the line of inheritance on a holding failed, through death or migration for example. Customary tenures held for life or a number of lives could be quite legally changed to leasehold at the end of the estate, since lords had no obligation to replace an estate for life with another estate for life. Lords could also buy out existing customary tenants at a price attractive to both lord and tenant, or sell a holding to a customary tenant and thereby convert it to freehold. In many cases the replacement of copyhold by leasehold did not make much difference to a tenant's standing with his landlord.

From the sixteenth century, many copyholds were replaced by 'beneficial leases'. These could be for a period of years, or more usually for life or a number of lives, and the rent was paid by the combination of a large entry fine and small annual payments (the 'ancient rent'). The establishment of such leases enabled landlords to raise a lump sum (the entry fine) at the cost of a low annual rent which meant they were effectively borrowing money from their tenants. These leases also regularised the tenants' position and gave power to enforce the conditions of tenure. For the tenant the beneficial lease removed the encumbrances of customary tenure. Generally landlords preferred rack rents to beneficial leases, since with estates for lives the timing of the major part of the rental income (the entry fine) depended on the unpredictable death of the tenant. By the later seventeenth century, rack rents were the usual mode of letting land in eastern England. Conversions to beneficial leases in the early sixteenth century had no adverse effects on tenants, but after mid-century, as prices began to rise, tenants recognised their advantage lay in hanging on to customary tenures, especially if entry fines were fixed. When local custom prevented lords from raising rents they could raise income by encouraging tenants to buy their freeholds; thus, in some areas, rather than the development of leasehold farms at rack rents, the sixteenth century saw the creation of more small freehold farms.

An example of this was in the village of Wigston Magna, just south of Leicester. Wigston had two manors, and one was sold off to customary tenants during 1586–8, probably because the lord's income was being squeezed between rising prices and fixed customary rents. Most of the customary tenants on this manor therefore became freeholders. By 1585 the other manor was owned by Sir John Danvers who was in dispute with his tenants over the nature of copyhold. The situation was confused because the manor had changed hands several times in the preceding years. The tenants argued that the custom of the manor gave them estates in fee simple (that is, of inheritance) with fixed entry fines; in practice this was little different from freehold. Danvers countered by saying that they only had an estate for one life and that he had the right to vary entry fines. Arable land on this manor was let at the fixed rent of 6d an acre – the level prevailing in the late thirteenth century – so the issue of the variability of entry fines was crucial to both lord and tenants. The outcome of the legal battle is unknown (although it seems likely that Danvers lost) but in 1606 the lands of the manor were sold to the tenants marking the end of the manorial system at Wigston. The open-field system with its common rights remained, but henceforward the village became a 'peasant village' in the sense that it had no large dominating landowner, and the field system was organised through the village rather than the manor.

In some cases the creation of leaseholds may have been clearly against the interests of customary tenants. New farms carved out of the waste would result in the removal of common rights and would invariably be let as leasehold farms. This process was quite frequent in the sixteenth century, and again in the late eighteenth and early nineteenth centuries when carried out by act of parliament. A second method of creating leaseholds was to manipulate entry fines (provided they were variable) by forcing them up to such levels that eventually customary tenants would be driven out. There is abundant evidence that entry fines rose dramatically in the seventeenth century but at the same time the law stipulated that such fines were supposed to be 'reasonable'. Finally, lords could have resorted to manifestly illegal activities by forcibly evicting customary tenants from the land.

It is possible to point to clear examples of landlords abusing their power and forcing tenants off the land or forcing them to switch to new leaseholds at rack rent. Such coercion could be downright illegal, or within the bounds of the law but involving unreasonable pressure and intimidation. Yet it seems, especially in the seventeenth century, that the law increasingly took the side of tenants in dispute with their lords. The law upheld manorial custom, and courts did arbitrate in disputes over the 'reasonableness' of entry fines. Of course lords with greater financial resources than their tenants might have been more able to withstand the rigours of a court case, but tenants were quite capable of taking their lord to court, and the examination of such cases does not suggest that the law always favoured the landlord.

By the late eighteenth century some areas of the country had a predominance of farms let on long leases at rack rent (for example Shropshire, Gloucestershire, Buckinghamshire and Kent), but a survey of the Board of Agriculture Reports has shown that the situation varied widely. Some counties (Cumberland, Worcestershire and Bedfordshire) had virtually no leases; while in other counties they seemed to be going out of fashion and were replaced by short-term annual tenancies. While a long lease encouraged tenants to make improvements to their holding, landlords could find it difficult to raise the rent during the currency of the lease. When prices were fluctuating, as they were during the Napoleonic Wars for example, annual tenancies minimised the risks of an uncertain future for the landlord.

Changes in property rights were usually associated with enclosure, but there were a number of legal devices for removing common rights. The first of these was termed the 'extinction of common in the ordinary process of law'. Under certain circumstances it was quite straightforward for common rights to be legally removed without any special procedures. They could be removed if the right was not exercised, or if the resource over which rights extended was no longer there (for example if an area of woodland had been cleared). A more frequent method of removal was when 'unity of possession' existed so that both the land and the common rights over the land belonged to the same person. Thus an individual could buy up a piece of land and the common rights over it, extinguish common rights, and enclose the land. Enclosure by unity of possession was most prevalent in the early sixteenth century.

If unity of possession did not exist, common rights could be removed if all those with such rights agreed to their removal. This is known as 'enclosure by agreement' and was the most frequent method of removing common rights. The 'agreement' could be a genuine voluntary agreement between those with rights to the land concerned but in some instances the consent of some recalcitrant parties was only achieved by hard pressure.

Enclosures by agreement were often ratified by collusive actions in the Court of Chancery in the form of fictitious disputes to enable the legal recognition of changes which had been made. Enclosure by agreement was primarily a seventeenth-century phenomenon, but continued well into the eighteenth century during the era of parliamentary enclosure. As with the abolition of copyholds, historians disagree over the extent to which enclosure by agreement really meant a genuine agreement of all those with an interest in the common rights of the village. Again it is possible to point to examples where landlords were ruthless in going against the wishes of the majority, but equally it is possible to point to examples of villagers taking their landlord to court.

One such example comes from the manor of Cotesbach, in Leicestershire, bought in 1596 by a London linen merchant, John Quarles. His tenants' leases expired in 1602 and he bargained with them for new leases at higher rents. When they rejected his offer he prepared to enclose the open-field to raise rents to a level comparable with the purchase price he paid for the manor. In this case the main obstacles came not from the customary tenants (who could legally be evicted at the termination of their leases) but from the freeholders in the open-fields whose customary rights would be infringed by enclosure. It was an expensive business to buy them out. The rector of the village held out against enclosure until Quarles agreed to pay the costs of enclosing his glebe. The tenants then petitioned the new King, James I, to stay the proposed enclosure but a commission of enquiry found for Quarles and he received a royal licence to enclose. This he did in 1603, and, although he offered his tenants new agreements, by 1607 sixteen farm houses had been abandoned and 520 acres of arable converted to pasture. Quarles was then brought before the Court of Star Chamber for depopulating the manor. On the eve of enclosure the yearly value of the manor was £300; after enclosure the manor was valued at £500 and the enclosure cost £500, although Quarles had laid out considerable sums in buying the manor and freeholders' farms. In the end his method of enclosing was by unity of possession.

A similar example is provided by Middle Claydon in Buckinghamshire, owned by the Verney family. By 1625 they had eliminated the small freeholds in the village by buying them out. They turned the copyholds into beneficial leases for three lives or 99 years, but all the new leases included a clause giving the Verneys the right to enclose and exchange the land for equivalent elsewhere in the parish. Thus the Verneys were able to enclose by unity of possession, although enclosure was gradual and piecemeal. By the seventeenth century all the demesne was enclosed, and in 1613 the demesne was extended to an area of wasteland over which common rights were extinguished. Woodlands and some open-field lands were enclosed in 1621 and the commoners were compensated by having the length of their estates doubled. Further enclosure followed in 1635–6 leaving the open-fields at no more than 500 acres. The commonfield system therefore had to be reorganised but by now it was too small to function as a viabie system. The next enclosure, in 1653–5, was therefore a general one. It is interesting to note that while the Verneys were actively enclosing Middle Claydon, they went out of their way to prevent the enclosure of the adjoining parish of Steeple Claydon. They were worried that the rents of their enclosed lands would fall if more enclosed land came on to the local land market. By buying a few acres and the right to graze cows on the common, they prevented the Challoner family from gaining unity of possession in Steeple Claydon, and delayed the enclosure of the village for 120 years.

The removal of common rights in the fenland did involve coercion, with royal authority. In the seventeenth century, the maintenance of drainage in the fenlands was under the authority of the Court of Sewers, who appointed commissioners to oversee and maintain drainage schemes. Royal advisers came up with an ingenious way of removing common rights as an obstacle to drainage. The commissioners were given the power to levy a tax on land 'hurtfully surrounded' by water. As a rule the tax was not paid, whereupon the commissioners seized the land and sold it to an 'undertaker' who would then drain the land in return for an allotment of newly drained land enjoying the full rights of private property.

From the mid-eighteenth century the most usual way in which common rights were removed was through a specific act of parliament for the enclosure of a particular locality. Such acts provided a sound legal basis to enclosure without the necessity of a fictitious suit in the Court of Chancery, but they also made the process easier because enclosure could be secured provided the owners of a majority (four-fifths) of the land, the lord of the manor, and the owner of the tithe agreed that it should take place. Thus the law of parliament (statute law) only took account of the wishes of those *owning* land as opposed to the common law which took account of all those who had both ownership rights and *use* rights to land. Moreover the majority required for enclosure was calculated in terms of acres rather than landowners, so in some parishes the four fifths majority could be held by a single landowner. Since the distribution of landholding in most villages was skewed, with a minority of owners in a village owning the majority of the land, parliamentary enclosure often resulted in a minority of owners in a village imposing their will on the majority of farmers. Those inhabitants owning no land but only enjoying common rights had no say at all, although they had the opportunity to petition against the enclosure bill. Once the act had passed through both Houses of Parliament and received the Royal Assent, Enclosure Commissioners appointed a clerk and surveyor to implement the provisions of the act on the ground.

Parliamentary enclosures removed common rights but they could also transform the agricultural landscape almost overnight. Subdivided fields were swept away and replaced with a new landscape of consolidated ring-fenced farms. New farmhouses were built in the centre of the new farms, altering the settlement pattern of the village away from a nucleated pattern to one that was more dispersed. New roads were often laid out in conjunction with the new fields, and many of the former paths and trackways over the commonfields were removed. As the Board of Agriculture Reporter for Lincolnshire put it in 1794: 'The first great benefit resulting from an enclosure is contiguity, and the more square the allotments are made, and the more central the buildings are placed, the more advantages are derived to the proprietors in every respect.' While wholesale enclosure by parliamentary act could result in dramatic remodelling of a village, gradual, piecemeal enclosures could work a similar transformation over a much longer period of time. Amalgamations of strips in the open fields, exchanges of strips between owners, and the enclosure of small groups of strips, gradually led to more efficient farm layouts. Once common rights had been removed there was little to prevent exchanges and amalgamations of land leading to 'ring-fence' farms. Fences were usually constructed at the time of enclosure in conjunction with hawthorn quicksets, which eventually took over the role as barriers to stock. Particular boundaries were associated with particular landscapes, so that stone walls were a feature of upland areas such as the Pennines, and ditches of low lying and ill-drained areas. Such changes did not meet with universal approval; the poet John Clare's vision of the new enclosed landscape was,

> Fence meeting fence in owners little bounds
> Of field and meadow, large as garden-grounds,
> In little parcels little minds to please,
> With men and flocks imprisoned, ill at ease.

From the fifteenth century onwards one of the main objectives in enclosing was to convert land from permanent arable in subdivided commonfields to permanent pasture in severalty. The rent of pasture land was consistently higher than arable, and, as has already been pointed out, livestock farming offered more possibilities for commercial agriculture, especially before the eighteenth century. Enclosed land gave more flexibility for farm production, be it for livestock, grain or a combination of the two, compared with commonfield farming. Periods of enclosure involving a change from arable to pasture are likely to coincide with periods when the price of livestock products rose relative to grain prices.

The conversion of arable to pasture in the sixteenth century aroused hostility from the state because of the alleged depopulating effects of such enclosures. The Husbandry Act of 1489 made it an offence to cause the decay of a farm with 20 or more acres, a 'house of husbandry'. In 1515, an act made it illegal to convert tillage to pasture; two acts in 1597 were also directed against engrossment and the conversion of land to pasture. Inquisitions were established to gather material for prosecutions under these acts, as in 1517 for example. In fact it is possible to identify two types of enclosure during this period: eviction enclosure, where the landlords' interest would be served by conversion to pasture for the reasons just mentioned, and abandonment enclosure, where enclosure followed the abandonment of a field system by its farmers. Population was still declining towards the end of the fifteenth century and villages were being abandoned when farmers moved away, perhaps to farm more fertile soils, leaving a non-viable commonfield system. In this situation the lord was forced to take over the commonfields and, in the absence of labour to work them, convert them to pasture. From the mid-sixteenth century prices no longer favoured livestock products and most subsequent enclosures were for arable farming. Growing commercial pressures encouraged enclosure for more intensive arable farming, such as in the light soil areas of the Chilterns, but it was also the case that in other areas, for example in Northamptonshire, the combination of market opportunities and the natural environment still meant that more profit was to be had in conversion to pasture.

A switch to livestock farming thus often followed when enclosure involved the removal of common rights and the engrossment of farms. The physical enclosure was also important since small enclosed fields would assist the management of livestock and prevent their 'promiscuous mingling' which was inimical to selective breeding. Enclosure for livestock farming was also important during the first phase of parliamentary enclosure (1760–80), when the worn-out cereal lands of many Midland parishes were transformed into permanent pasture. Often fields laid down to pasture at this time have not been ploughed since, leaving the ridge and furrow pattern of the open-field strips fossilised in the present landscape.

A change from pastoral to arable farming was most common with 'reclamation enclosure' which resulted in the intensification of land use from poor-quality pasture to arable. This intensive farming has been discussed in the previous chapter, and was particularly associated with land reclamation in the second phase of parliamentary enclosure (1790–1820), although it had started before then. The contrast in types of enclosure is indicated by Table 4.6 which shows the situation before and after the parliamentary enclosure of four categories of land. The data come from a work promoting the enclosure of wastes (*The advantages and disadvantages of enclosing waste lands, by a country gentleman*) so it might well be biased in favour of reclamation enclosure. Even so, the table suggests that increased profits and rents from the enclosure of 'Rich open-fields', with an

Table 4.6 *Costs and benefits of four types of parliamentary enclosure (£ per 1000 acres)*

Type	Wool	Produce	Total	Labour costs	Horses	Other costs	Rent	Profit
A. Rich open-fields								
Open	50	2350	2400	400	367	966	300	364
Enclosed	250	1250	1500	100	25	125	750	500
B. Poorer open-fields								
Open	50	1950	2000	400	367	733	200	300
Enclosed	100	1700	1800	325	250	455	400	370
C. Rich common pastures								
Open	100	370	470	10	0	120	100	240
Enclosed	250	1250	1500	100	25	125	750	500
D. Commons, heaths and moors								
Open	90	100	190	10	0	70	50	60
Enclosed	100	1700	1800	325	250	155	400	370

associated land use change from arable to pasture, resulted in a fall in the value of output which was more than compensated for by a fall in production costs. The enclosure of 'Poorer open-fields' also saw a fall in the value of sales, but a reduction in labour and traction costs that enabled profits to rise slightly. What is especially noticeable here is that although profits only rose by 23 per cent (from £300 to £370 per acre), rent per acre rose by 100 per cent (doubling from £200 to £400), implying a greater gain for the landlord than for the farmer.

Enclosure of the third type of land in Table 4.6, 'Rich common pastures', represents a move towards arable production, with a threefold increase in output, a sevenfold increase in rents and a doubling of farm profits. But the most spectacular transformation takes place with the final category of land, 'Commons, heaths and moors', where intensive arable production, presumably using techniques such as the Norfolk four-course rotation, resulted in a rise in sales and rents of over eightfold, and a sixfold rise in profits. We do not know exactly how these four types of land were defined, but Figures 2.6 and 2.10 give some indication of the location of the lightland 'Commons, heaths and moors', in the areas of parliamentary enclosure shown in Figure 4.3.

The economic causes and consequences of enclosure

This discussion of changes in tenures, property rights, topography and land use has implicitly suggested some motives for enclosing. Just as there is no shortage of theories to explain the origin of commonfield systems, so there has been much discussion among historians as to why enclosure took place when it did. The causes and consequences of enclosure are considered together here since the motives for enclosure were obviously related to its perceived effects, although some of the effects of enclosure would not have had much influence on motives for enclosing, and, since people cannot see into the future, there are likely to have been unintended and unforeseen consequences of the enclosure process. Many of the arguments about enclosure have been concentrated on the period since the eighteenth century, since documentary material on parliamentary enclosure is so readily available, but the following discussion applies to enclosure from the sixteenth century.

The main incentive for landlords to enclose was that enclosed land was worth more than open commonfield land. This of course was most attractive to the owners of land whose principal income came in the form of rents. The magnitude of the increase in rents on enclosure is fairly easy to demonstrate. At Middle Claydon, for example, the rent income was between £1,000 and £1,400 before the mid-seventeenth-century enclosure, but over £300 more afterwards, representing a return on the direct costs of enclosure of between 50 and 100 per cent. There are countless other examples to show that several land was worth more than common land, that an open-field farm was worth less than the same-sized farm when enclosed and ring-fenced. William Marshall, the eighteenth-century agricultural writer, considered that land intermixed in open-fields or common meadow was worth one third less than equivalent enclosed land, and if common of shack existed on the open-fields it reduced their value to one half of the equivalent enclosed land. The general consensus has been that rents doubled with enclosure, but recent work suggests that this might be an exaggeration and that the average rent increase was nearer 30 per cent.

While it is clear that enclosure raised rents, it is much less clear as to the source of these increased rents. Even if rents doubled it does not follow that farmers' profits doubled, still less that land productivity doubled. If output did increase with enclosure, and rent remained the same proportion of output, then the proportionate increase in rent would reflect the proportionate increase in output. But it is unlikely that this was the case. Farmers' profits are a function of costs as well as income, and enclosure could have resulted in a reduction in costs if farmers saved labour. The proportion of profits taken as rent from tenants by landlords is the outcome of a power struggle between the two groups, and the increase in rent with enclosure may simply reflect an increase in landlord power. When enclosure took place during a period of inflation (as in the late eighteenth century) new leases created the opportunity to replace rents that had been fixed when the price level was lower. This represents a transfer payment from farmers to landlords, which also took place in earlier centuries when enclosure presented the opportunity to switch from copyholds with fixed fines and low rents, to leases for a term of years at rack rent. Enclosures of the waste could involve the largest transfer payments of all, since the owner of the waste now had the sole rights to land that been used by many people. Care must also be taken in making comparisons of rent before and after enclosure. Tithes were often commuted at enclosure, so land would be worth more after enclosure because it was tithe-free. Land use often changed with enclosure, enabling more profits to be made from alternative enterprises and therefore more potential rent for the landlord. Although work has been carried out on the costs of enclosure it is difficult to calculate the rates of return, especially as perceived by a landlord contemplating enclosure. Thus attempts to explain the incidence of enclosing activity specifically through fluctuations in the rate of interest or movements in agricultural prices are rather hazardous.

The picture is further complicated by the local histories of enclosure. Often, the key to variations in the rate of enclosure lay in the pattern of landholding which could vary dramatically from parish to parish. Resident lords, owning most or all of a village or township, were usually able to enclose with relative ease, whereas communities of freeholders found it more difficult to reach unanimity and agree upon enclosure. Enclosure in such parishes was more likely when the benefits of enclosing became obvious to all, for example during the inflationary period of the Napoleonic Wars.

It is clear that enclosure provided opportunities for greater profits from farming even though rents may be an inadequate guide to the magnitude of the increase. The type of farming could be changed almost overnight, new farming innovations could be introduced without hindrance, costs could be reduced through employing less labour, and the quality of land could be improved through capital investment in reclamation and drainage. The commonfield system was devoted to the needs of subsistence grain farming – to the maintenance of corn-growing through the careful use of animal resources. The scattering of strips could therefore have been a device to prevent enclosure: scattered strips (and common rights) forced individuals to follow the husbandry course of the village and ensured arable husbandry was maintained. When the imperative for such subsistence farming in particular localities was removed – through gradual rises in output, improvements in marketing, and the benefits of specialisation – then the counter force of the market became the controlling influence, stimulating the desire for the profits to be gained from enclosure. The commonfield system also ensured that small farmers enjoyed the benefits of economies of scale available to large farmers, in particular from the folding of a large sheep flock on their land. As farm sizes increased, especially in the eighteenth century, more individual farmers were able to enjoy such economies and enclosed farms became more attractive. It has also been suggested that strips were scattered to minimise the risk of harvest failure to an individual farmer. If the variance or range of yields from strip to strip in a village was high then an individual farmer was more likely to achieve the average yield for a particular year if his crop was scattered around the village. As the variance of yields in a village reduced, and as farms got larger and could carry stocks of grain from year to year as an insurance, the benefits of scattering diminished, and enclosure became a more attractive option.

Historians have tackled the question of the impact of enclosure on output and productivity in two ways: first, using evidence claiming to show husbandry changes following enclosure, and second, using evidence claiming to show how grain yields changed following enclosure. In a study of Huntingdon and Rutland, based on surveys by the Board of Agriculture Reporter, there was significantly more innovation of turnips and clover in enclosed villages (in the light-soil districts) than in commonfield villages; a finding reinforced by evidence from the tithe files from thirty-three light-soil villages. Enclosed villages had 3 per cent of their arable under fallow, but 23 per cent under clover and 20 per cent under turnips. Of the fifteen open-field villages six grew no turnips and had an average of 24.8 per cent of their land in fallow, and, on the other nine that were growing turnips, the turnip proportion averaged 14 per cent with 11 per cent of arable land as fallow. In the heavier soil areas of the South Midlands, enclosed villages were associated with extensive land improvement through draining.

In fact, strictly speaking this evidence is not showing the situation before and after enclosure, but is cross-sectional data, showing the situation in both open and enclosed parishes at the same point in time. A number of studies have employed a similar cross-sectional methodology comparing yields in open-field parishes and enclosed parishes. Turner, using the 1801 crop returns, shows a 25 per cent difference in wheat yields between open and enclosed parishes, but according to Allen does not control adequately for differences in soil type between parishes. In the South Midlands, Allen finds a difference of only 3 per cent for yields of wheat; although for barley it is 19 per cent and for oats 39 per cent. These studies illustrate the distortions that can arise by concentrating on wheat yields alone: although rises in wheat yields were modest, most of the enclosure that had taken place by this time had been for livestock rather than grain farming, so wheat is not the most appropriate indicator of increased land productivity. But the major problem with these cross-sectional studies is that a comparison between enclosed and open villages is particularly inappropriate at this time because the nature of enclosure changed. The first wave of parliamentary enclosure was often associated with the switch of worn-out arable land to grass, and grain yields in these cases might well have been poor. Many of the remaining open villages are likely to have remained open because their productivity was relatively high, thus minimising the difference in yields between enclosed and open villages.

To measure the direct effects of enclosure in a particular parish we need information about conditions in that parish both before and after enclosure. Fortunately such information exists in a survey of the clergy carried out by the Board of Agriculture. The results of this survey were published in terms of the number of enclosures resulting in an increase or a decrease in various elements of farm enterprises. The survey shows that the acreage of wheat reduced with enclosure, but that the acreages of barley and oats increased. Numbers of sheep and cattle rose dramatically and the clear implication is that land productivity was rising too.

Particularly detailed information of the situation before and after enclosure is available for Canwick near Lincoln, and although merely an example of one parish (of a light-land 'reclamation enclosure') it is instructive. Table 4.7 shows that following enclosure wheat yields rose by only 10 per cent, from 20 to 22 bushels per acre, but that barley and oat yields rose by 40 and 78 per cent respectively. The most significant change, however, was with livestock; the numbers of sheep rose by 33 per cent but the value of their output increased by an astonishing 590 per cent, from £200 for wool alone before enclosure to £1,380 for mutton and wool after enclosure. This was because flocks kept for folding on the arable commonfields were only valued for their wool, but were replaced at enclosure by flocks of improved breeds of sheep kept primarily for their mutton and fed to a much higher level with fodder crops, in this case clover. This example also indicates how misleading wheat yields can be as an indicator of land productivity since they only rose by 10 per cent. The rental value of the village was £730 before enclosure (in 1760), but after enclosure had almost doubled to £1,380 in 1790, and during the war years doubled again, reaching £3,200 in 1812.

Enclosure facilitated innovation and changes in land use because the constraints imposed by common property rights, the scattering of land, and collective decision making could be overcome. Contemporaries were virtually unanimous that enclosed fields offered more opportunities for making money than did commonfields. Nathanial Kent put the arguments as follows:

Table 4.7 *Farming before and after enclosure in 1786 at Canwick, Lincolnshire*

	Open-field acres	Enclosed acres	% Change	Open-field yield (bushels)	Enclosed yield (bushels)	% Change	Open-field output (bushels)	Enclosed output (bushels)	% Change
Wheat	228	150	-34	20	22	10	4560	3296	-28
Barley	436	380	-18	20	28	40	8720	10640	18
Oats	48	40	-20	27	48	78	1280	1920	50
Peas	80	60	-33	18	28	56	1440	1680	17
Tares	100	0							
Turnips	218	250	15						
Clover seeds	0	593							
Fallow	300	0							
Total crops	1410	1473	5						
	No.	No.					Cwt	Cwt	
Animal food							447	981	120
							£	£	
Bullocks	60	64	7				480	378	-21
Cows	69	34	-51				414	204	-51
Sheep (mutton)	0	580					0	980	
Sheep (wool)	1200	1600	33				200	400	100
Horses	73	53	-27						
Total livestock products							1094	1962	79
Total cereals sold							2787	2692	-4
Total produce							3881	4654	20

Land, when very much divided, occasions considerable loss of time to the occupier, in going over a great deal of useless space, in keeping a communication with the different pieces. As it lies generally in long narrow slips, it is but seldom it can receive any benefit from cross ploughing and harrowing, therefore it cannot be kept so clear; but what is still worse, there can be but little variety observed in the system of cropping; because the right which every parishioner has of commonage over the field, a great part of the year, prevents the sowing of turnips, clover, or other grass seeds, and consequently cramps a farmer in the stock which he would otherwise keep. On the contrary, when land is inclosed, so as to admit of sowing turnips and seeds, which have an improving and meliorating tendency, the same soil will, in the course of a few years, make nearly double the return it did before, to say nothing of the wonderful improvements which sometimes result from a loam or clay; which will, when well laid down, often become twice the permanent value in pasture, that ever it would as ploughed ground.

In short, the farmer on enclosed land, in Kalm's words, 'could in a thousand ways improve his property and earn money'.

When land was under private property rights the return on investment made in that land by an individual would accrue to that individual and not to the community as a whole. Drainage, liming and marling, floating meadows and other capital improvements were therefore much more likely with the removal of common rights. Farmers sowing turnips on their commonfield strips would see them eaten by other people's livestock in the autumn if they exercised their right to the common of shack. Selective breeding of livestock was more difficult if animals mingled on the commons, whereas small enclosures gave more opportunity to manage animals more efficiently. This is not to say that enclosure inevitably led to better farming: 'severalty makes a good farmer better but a bad one worse', and the regulation of commonfield systems improved the farming of the worst farmers as well as holding back the more able. Some historians have argued that open-fields were not as inflexible as these contemporary opinions suggest. Evidence has been found of commonfield communities deciding to grow grass substitutes such as clover and sainfoin on their fallow field. Turnips, too, were grown in the open-fields, and farmers modified their bye-laws to accommodate them. But, despite these examples of innovation in commonfields, enclosure accelerated the process dramatically and gave immediate opportunities to make new profits, and the transformed landscapes it produced were a constant reminder that a new agricultural order was in place. A major conclusion of the previous chapter was that the major upsurge in agricultural output and productivity came after the mid-eighteenth century: this coincides with the major burst of parliamentary enclosure.

Table 4.8 *The distribution of landownership in England and Wales, 1436–1873 (percentages)*

	1436 (England)	c.1690	c.1790	1873 (England)
Great owners	15–20	15–20	20–25	24
Gentry	25	45–50	50	55
Yeomen freeholders	20	25–33	15	10
Church	20–30			
		5–10	10	10
Crown	5			

The changing social structure: social differentiation

Changes in the ways in which land was held, the move from communal to individualistic farming, the reorganisation of farms and the establishment of private property rights were radical changes in the conditions under which farming was carried out. They also contributed to changes in the fortunes of the various social groups in the countryside, particularly in their relationships to land. In addition to these factors, however, the changing economic environment of farming, the movements of costs and prices, became ever more important as more and more farmers became involved in commercial farming. The broad outline of changes in land ownership in England from the fifteenth to the nineteenth century is shown in Table 4.8. The percentages in the table must be regarded as indicative rather than strictly accurate, and the groups they refer to are fairly loosely defined and may vary slightly from period to period. The table shows a massive decline in the amount of land owned by the church; a rise in the landholding by the gentry; a small rise in the proportion of land held by the great owners in the eighteenth century; and a decline in the proportion of land owned by the yeomen freeholders. These trends form the first four themes for this section of the chapter: the plunder of the church, the rise of the gentry, the rise of the great estates and the decline of the small farmer. The final two themes are related to the decline of the small farmer, and are the rise of an agricultural proletariat and the decline of servants.

'The plunder of the church' is Hoskins' description of what others call the dissolution of the monasteries, when Henry VIII seized monastic property and lands. In the 1530s the lands of the church were yielding an income in the region of £400,000 (when the crown lands were providing an income of only about £40,000) and about half this wealth came from monastic houses, as the direct profits of manors, and also from tithe and other income. From 1536, at least 60 per cent of this wealth was transferred to

Table 4.9 *The ownership of manors in Norfolk, 1535–65 (percentages)*

Date	Crown	Nobility	Gentry	Church	Monasteries
1535	3	10	67	3	18
1545	8	13	73	6	0
1555	5	12	77	7	0
1565	5	11	78	6	0

the crown. The lands passed through the crown's hands quite quickly and into the possession of many smaller landowners, particularly the gentry. The lesser monastic houses were suppressed and their lands confiscated in 1536, and in 1539 the larger houses were similarly treated. In the former group came about 374 monasteries worth some £32,000 annually. There were over 180 greater monasteries with land bringing in some £100,000 per annum in addition to the capital value of plundered treasures. The plunder continued, so that in 1545 the property of various colleges, chantries, chapels and hospitals passed to the crown, and, by an act of 1542, some 700 Irish monasteries. The vast majority of lands taken from the church were sold relatively quickly, as the example from Norfolk, shown in Table 4.9, demonstrates. Local studies suggest that the land went to established local families: mostly to the gentry both old and new, although the new were often younger sons of existing gentry families. Favourites of the crown also benefited, and the contemporary consensus was that the episcopal estates seized by the crown wound up in the hands of the courtiers and their friends.

In 1941 Tawney published his thesis of the 'rise of the gentry': that in the century before the civil war, a new bourgeoisie or capitalist class of farmers emerged while the old aristocracy declined, and that the changing balance of economic power led to a change in political power which was to be reflected in the English revolution. Tawney's ideas were challenged by Trevor-Roper who argued that the distinction between gentry and nobility was not the important one since both groups faced the same economic pressures. He claimed the more important distinction was between those who succeeded in prospering through office-holding and those who did not – the latter being described as the 'mere' gentry. Nevertheless, it seems clear that the gentry class (as described in Chapter Two) did grow considerably in numbers from the mid-sixteenth century, since the evidence in Table 4.8 is reinforced by many local studies of landholding.

While the active land market prompted by the dissolution of the monasteries fuelled the transfer of land to the gentry, a more important dynamic was the general inflation of the sixteenth century. Food prices (shown in Table 3.1 and Figures 3.2 and 3.3) were rising more rapidly than the price of industrial products or such consumer goods as there were at the time. This made holding land attractive, and encouraged the rise in rents which seems to have outstripped the rise in agricultural prices. Thus, while farmers benefited from inflation as a rule, landlords benefited to a greater extent, although there were considerable variations depending on individual circumstances.

The civil war caused a severe, if probably short-term, disruption to land-lords, as it did to many farmers. There was no revolution in landholding, and those who had supported the Royalist cause were the group to suffer most. They were fined, had their lands confiscated, and were subject to high levels of taxation. Although most Royalists managed to get their lands back, they were likely to be encumbered with debts. According to the contemporary John Houghton this had a beneficial effect on agricultural production, since he speaks of, 'the great improvement made of lands since our inhuman civil wars, when our gentry, who before hardly knew what it was to think ... fell to such an industry and caused such an improvement, as England never knew before'. Price inflation came to an end in the mid-seventeenth century and for several decades rents and land prices languished. In these more adverse economic conditions it would seem likely that larger landowners fared better than smaller ones. They had more security to borrow money, and were more likely to get money from office. Holding land also provided a source of political power and social prestige so that the advantages of landholding went beyond the revenue from rents. While there were pressures by the larger landowners to purchase more land, those at the other end of the scale were facing pressures encouraging them to sell. The argument therefore is that the century after 1660 witnessed the build-up of large estates owing to pressures on both the demand for and the supply of land. Inheritance practices and marriage settlements limited the supply of large estates coming onto the market. The device of the 'strict settlement' enabled land to be held in trust from generation to generation and prevented it from being sold or broken up into smaller parcels. This created a shortage so that acquiring land through marriage became more important. Smaller owners, on the other hand, were prompted by low profits caused by low prices to sell up and leave the land. As a result of these processes Habakkuk argues, using evidence from Northamptonshire and Bedfordshire, that the general drift of property in the sixty years after 1690 was in favour of the large estate and the great lord.

This thesis has recently received some revision: more recent research has suggested that the build-up of estates in some areas was matched by their dissolution in others. In Lincolnshire, for example, there is less evidence for the growth of large estates, and generally the land market seems to have been more fluid than Habakkuk's findings suggest. Strict settlement seems to have been of exaggerated significance because trusts could be broken, but was nevertheless an important device for preserving large estates, although settlements made by will were more important than those made in marriage settlements. The development of the modern law of mortgage also increased the propensity of lords to buy up freeholds and copyholds. From the 1660s onwards it became much easier to obtain a long-term loan to buy land, using existing land as security, whereas previously mortgages could only be held for a short time.

The argument also needs to be modified from the perspective of the supply of land. Economic pressures on small farmers in the century after 1660 might have encouraged those heavily engaged in commercial farming to sell up, but small farmers could insulate themselves from such market forces by retreating into self-sufficiency. Although prices were depressed in the century after 1660, the extent of the depression varied for different products, and the effects of low prices affected different categories of farmers in different ways. Larger grain farmers on the lighter sheep-corn lands were probably more resilient, since they had more opportunities to improve efficiency, could benefit from economies of scale and, by selling a larger volume of grain, had more bargaining power with the emerging corn merchants. Grain farmers on heavier land could switch their enterprise over to livestock to benefit from better prices. Livestock farmers enjoyed better prices than their contemporaries in arable areas (with the exception of wool prices) and were more likely to be able to supplement their income with by-employments. Those in the most precarious economic position were grain farmers, farming on a small scale, yet subject to the commercial pressures of the market.

The reduction in the number of small farmers is important because of its relationship to the build-up of large estates, but it is also an important issue in its own right. The increased polarisation of landholding and the diminution of small farms is an important theme in English rural history, and, in turn, is linked to the rise of a class of landless rural farm labourers, or the rural proletariat. This phenomenon goes under a number of labels, including 'the decline of the small landowner', 'the decline of the small farmer' and 'the disappearance of the peasantry'. This variety of labels suggests some ambiguity as to the exact nature of the issue being investigated and this is indeed the case. For this reason we must return to the concept of the peasantry introduced in Chapter Two.

Differing definitions of the peasantry result in different chronologies for their disappearance. At one extreme, using Macfarlane's eastern European definition, an English peasantry never existed. At the other extreme, adopting Mingay's argument, and turning the debate about the decline of the peasant into the fate of the 'small farmer' of less than 100 acres, then in 1870, as Table 4.13 shows, all but 18 per cent of English farmers were

Table 4.10 *Landholding in Chippenham and Willingham (percentages)*

Farm size (acres)	Chippenham		Willingham	
	1544	1712	1575	1720
> 250	0.0	6.1	0.0	0.0
90–250	3.0	8.1	1.0	0.0
45–90	19.7	4.1	0.0	1.3
15–45	18.2	2.0	32.0	17.0
2–15	6.1	6.1	21.0	37.3
< 2	21.2	10.2	8.0	13.1
Landless	31.8	63.1	38.0	31.4

small farmers. In the present context the fate of the 'peasant' will be examined from two points of view: first, through changes in the distribution of farm sizes, and second, in a wider context, through investigating whether or not a class of people living on the land with a distinctive way of life actually disappeared.

Several economic mechanisms had the effect of raising farm sizes for particular groups of farmers and forcing the smallest farmers out of business. We have already seen in Chapter Two (Table 2.2) that a poor harvest could leave subsistence farmers with insufficient food, so that they were forced to buy corn in the market when its price was high but had corn to sell when harvests were good and prices were low. Such bad harvests are reflected in the annual graph of wheat prices in Figure 3.2. Runs of particularly bad harvests occurred in 1501–2, 1520–1, 1550–1, 1555–6, 1594–7, 1650–1, 1660–1, 1673–4, 1691–3, 1696–7, 1708–9, 1739–40, 1756–7, 1795–6, 1799–1800, 1808–12 and 1816–17. A bad harvest, or worse, a run of bad harvests, brought this process into effect. It is evident for example in the village of Chippenham in Cambridgeshire, an arable village where the disappearance of the small landowner began in the late 1590s, took thirty years, and was caused primarily by price movements and dearth. Table 4.10 shows the landownership structure before and after this transformation. The table also shows the situation in the nearby fenland village of Willingham which had a completely different economy, and while there was some adjustment in farm size groupings, stock farming and dairying made it economically feasible for small farmers to survive.

Counterbalancing this process were other forces creating new small farms. This was especially the case in the sixteenth century, when, under conditions of population pressure and high land prices, land was more likely to become fragmented, especially in areas where partible inheritance predominated. New farms were also created by colonisation of the waste, particularly in upland areas. However, it is likely that small farmers

Table 4.11 *Farm sizes in the South Midlands*

| Acreage | Percentage by acreage | | | | | | Percentage of farms | | | | | |
| | open | | | enclosed | | | open | | | enclosed | | |
	(1)	(2)	(3)	(1)	(2)	(3)	(1)	(2)	(3)	(1)	(2)	(3)
5–15	1.8	2.2	0.2	0.6	1.0	0.1	11.9	15.0	2.7	13.0	8.3	1.8
15–30	5.8	6.8	1.2	0.5	2.5	1.8	16.2	20.1	6.9	4.3	13.1	11.9
30–60	25.9	15.7	6.2	5.3	9.1	6.2	34.8	24.1	20.0	26.1	21.4	21.1
60–100	34.2	21.7	7.6	2.8	15.1	6.4	25.6	18.3	15.2	8.7	19.0	11.9
100–200	21.9	41.3	28.3	9.6	43.9	25.5	9.8	19.3	26.2	17.4	29.8	24.8
200–300	3.6	7.7	30.8	13.5	11.5	26.5	0.9	2.3	19.3	13.0	4.8	15.9
> 300	6.8	4.6	25.6	67.8	16.8	33.9	0.9	0.8	9.6	17.3	3.6	12.6

(1) Early seventeenth century
(2) Early eighteenth century
(3) *c.*1800

continued to be forced out by dearth in the seventeenth century, especially during the run of five bad harvests in the 1690s. By this period however, commercial pressures were increasingly affecting small cereal farmers no longer farming at subsistence levels. In these conditions larger grain farms generally had lower costs per acre than did small farms; the capital needed for a 60 acre farm was probably the same as the amount needed for a 150 acre farm for example, and larger farms could secure economies of scale and practice division of labour. Economies of scale meant that large farmers made higher profits per acre than small farmers. This probably mattered little to small owner occupiers, but it would be a concern to landlords, provided rents per acre could be linked to the profits their tenants made: larger farms would therefore mean higher rents per acre, and landlords would have an incentive to engross farms to make larger holdings.

Eventually, by the eighteenth century, pressures of increased commercialism began to affect more and more farmers, and progressive and widespread increases in farm sizes took place. The most thorough evidence of this has been compiled by Allen and is shown in Table 4.11. His evidence shows a fairly continuous increase in farm size from the early seventeenth century although the eighteenth century was marked by dramatic increases in farm sizes on open-field land. Allen's findings are corroborated by other evidence of farm size. In four Nottinghamshire manors for example, in 1690, 93 per cent of open-field farms were under 100 acres and 7 per cent above; by 1790 the proportions were 53 and 47 per cent. Table 4.12 shows changes in farm sizes over the eighteenth century on the Leveson-Gower estates in Staffordshire, Shropshire and Yorkshire, and again, the picture is similar: large farms increase at the expense of small ones.

Table 4.12 *Farm sizes on the Leveson-Gower estates (percentage distributions by acreage)*

Date	0–20 acres	20–100 acres	100–200 acres	over 200 acres
1714–20	6.3	46.1	28.8	18.8
1759–79	6.2	26.6	35.0	32.2
1807–13	6.5	16.7	25.1	51.7
1829–33	9.6	14.9	16.2	59.3

National statistics of farm sizes were collected for the first time as part of the 1851 census, but the data are unreliable because the question was a voluntary one and many small farmers did not bother to give an answer. The first reliable nation-wide information is therefore from an inquiry attached to the agricultural statistics for 1870, shown in Table 4.13. The 1870 data have no information on farms above 100 acres so the 1851 figures have been used to show the percentages above 300 acres. These undoubtedly exaggerate the proportion of farms in this category although regional variations should be reflected more or less accurately. The largest farms in the country were associated with extensive grain growing, on the chalklands of Berkshire, Hampshire and Wiltshire for example.

It is noticeable that there is no direct relationship between the pattern of farm size and parliamentary enclosure, at least at a county scale: enclosure *per se* did not create large farms. In the South Midlands, as Table 4.11 shows, the magnitude of the decline was almost as great for open-field farmers as it was for those farming in severalty. On the Leveson-Gower estates (Table 4.12) the land was old-enclosed and so again the changes in farm size cannot be due to enclosure. Thus the decline of a 'peasantry' defined narrowly in terms of farm size alone seems to have been a fairly continuous process, accelerating from the mid-seventeenth century, but due more to commercial pressures than to enclosure and eviction. Some historians have now gone so far as to argue that England had no peasants by the second half of the eighteenth century.

Yet enclosure, and particularly parliamentary enclosure from *c.* 1750 to *c.* 1850, was held by an early generation of agricultural historians to be responsible for eliminating the peasantry. This brings us to the second way of looking at the decline of the 'peasantry'. Recent research is now claiming that enclosure did indeed contribute to the creation of a class of landless labourers and to the disappearance of a distinctive social class from the countryside. The Hammonds, writing in the early part of this century, were certain that enclosure was responsible for driving peasants off the land. After the Second World War, revisionist historians argued that enclosure did no harm to the small farmer and that the peasantry had already disap-

Table 4.13 *The distribution of English farm sizes in 1870*

	Total farms	Percentages in acreage groups					
		<5	≥5–<20	≥20–<50	≥50–<100	≥100	>300^a
Bedfordshire	3752	30	28	13	8	21	*16*
Berkshire	3927	28	25	12	9	26	*27*
Buckinghamshire	5548	32	21	12	10	25	*13*
Cambridgeshire	6715	27	29	14	10	20	*14*
Cheshire	13034	29	30	17	12	12	*1*
Cornwall	13542	30	30	18	12	10	*2*
Cumberland	7473	14	21	20	23	22	*5*
Derbyshire	12736	25	34	19	11	11	*2*
Devon	17326	20	23	18	18	20	*5*
Dorset	4802	25	25	15	11	24	*20*
Durham	6157	18	30	15	14	24	*7*
Essex	9381	22	21	14	14	29	*16*
Gloucestershire	10447	34	25	12	10	19	*12*
Hampshire	8434	33	25	12	8	22	*22*
Herefordshire	6701	29	26	13	10	21	*9*
Hertfordshire	4116	31	21	13	10	26	*21*
Huntingdonshire	2748	28	28	13	9	22	*18*
Kent	10319	19	28	17	13	22	*12*
Lancashire	21745	19	36	26	13	7	*1*
Leicestershire	8044	20	31	17	12	20	*7*
Lincolnshire	24518	26	32	15	9	17	*11*
Middlesex	2530	28	32	16	11	13	*10*
Monmouthshire	4512	22	28	19	16	15	*3*
Norfolk	16995	38	23	14	9	16	*13*
Northamptonshire	6721	20	25	15	12	28	*17*
Northumberland	5497	20	23	12	10	35	*25*
Nottinghamshire	8265	27	34	14	9	16	*7*
Oxfordshire	4515	26	22	12	11	30	*16*
Rutland	1369	20	32	17	12	19	*1*
Shropshire	11198	30	28	12	9	21	*9*
Somerset	14942	27	27	16	13	17	*5*
Staffordshire	12895	30	31	15	10	13	*5*
Suffolk	9328	26	19	15	14	26	*11*
Surrey	5153	28	29	15	11	18	*12*
Sussex	8492	20	25	18	14	23	*14*
Warwickshire	7432	25	28	14	10	22	*10*
Westmorland	3623	11	24	25	23	18	*5*
Wiltshire	7633	35	22	11	8	24	*21*
Worcestershire	6975	30	28	14	11	18	*8*
Yorkshire East	8382	29	24	12	10	25	*12*
Yorkshire North	14797	27	26	15	13	18	*5*
Yorkshire West	30850	24	38	18	9	10	*2*
England	393569	26	28	16	12	18	*10*

^{a} The percentage over 300 acres is for 1851

peared. Chambers considered that enclosure resulted in more employment not less, because of the extra work of hedging and ditching at the time of enclosure and the subsequent adoption of labour-intensive fodder crops. Evidence of the number of owners recorded in the land tax and the amounts they were paying in tax was also used to show that the number of small owners appeared to increase rather than diminish with enclosure. This revisionist view is now itself under revision. As Table 4.6 suggests, some types of enclosure could result in increased employment but they were balanced by those that did not. The national evidence of labour productivity suggests that enclosure was not leading to proportionately greater employment in the countryside. The land tax evidence is usable only after 1780, by which time many enclosures for pasture had taken place and reclamation enclosures, which might have increased employment, were the most prevalent. The evidence is the subject of considerable debate and the relationships between the number of taxpayers and the number of owners, and between the amounts of tax paid and the acreages of land held, are problematic. However, by looking at the turnover of names in the land tax documents at enclosure, it is possible to observe radical changes in the landowning structure. Parishes in Buckinghamshire and Northamptonshire showed about a 40 per cent turnover in landowners after enclosure compared with around 25 per cent for the same period in parishes either remaining open or already enclosed. Moreover, small farmers (with between 5 and 25 acres), whether they were tenants or owners, were more likely to give up their lands following enclosure than larger farmers. In fact the number of owners could increase at enclosure as holders of common rights were compensated by small allotments of land. These were often too small to make a viable farm and were often sold within two or three years of the enclosure being enacted. Neeson's study of Northamptonshire shows that enclosure witnessed a high rate of turnover of landholding, a striking contraction in the size of original holdings, and an absolute decline in the number of small owner-occupiers, landlords and tenants. Small farmers had to sell land to finance enclosure costs, and they also lost about 20 per cent of their land in lieu of tithes.

Thus, notwithstanding the increase in farm sizes from 1660, by the mid-eighteenth century there were still substantial numbers of very small farmers and commoners. Their proportion of the cultivated acreage was small, but they represented a sizeable minority, and a very distinctive group, in the countryside. Many of these 'farmers' were farming plots too small to provide for their subsistence, and were dependent for their livelihood on common rights. These commoners subsisted through a variety of means. They might own a little land which they cultivated for food; they might work for other people from time to time, but not necessarily on a regular basis; they might be involved in some handicraft activity, but again, not on a full-time basis and only when they chose; and they enjoyed the benefits of common rights. These could be quite substantial. A cow, for example, supported by the grazing on the common, could provide an income equivalent to half the annual wage of a labourer. The perquisite of gleaning could provide flour for a family for several months. Many of these commoners could not be called farmers because their landholdings were too small for subsistence, but nor were they labourers, because they worked for others only occasionally. The removal of common rights could undermine their independence by unravelling the complex framework of survival and force them to work for others on a full-time basis as their sole source of income.

Although the *owners* of common rights received some compensation at the time of enclosure, compensation was not paid to those tenants *enjoying* common rights because they were leased (along with a cottage for example). In some areas there were also groups of people enjoying the benefits of common land without any legal right to those benefits. Many commons were inhabited by squatters, and one motive for enclosure was to squeeze out these unwelcome groups. For example, in 1777 the enclosure act of Ipstone in Staffordshire stated: 'It will put a stop to many encroachments that are every day making upon the commons by people who have no right to them and will keep many bad people out of the neighbourhood.' Eighteenth-century propagandists in favour of enclosure had no doubt of the effects of enclosure on commoners and the smallest farmers. According to Arthur Young,

There is, however, one class of farmers which have undoubtedly suffered by enclosures; for they have been greatly lessened in number: these are the little farmers ... That it is a great hardship, suddenly to turn several, perhaps many of these poor men, out of their business, and reduce them to be day-labourers, would be idle to deny; it is an evil to them, which is to be regretted.

This conclusion was based not merely on his own observations, but on the evidence from returns from the clergy on the effects of enclosures. In most counties, enclosure resulted in a fall in the number of cows, as the commoners lost their rights (an example is shown in Table 4.7). In addition to the loss of common rights, the smallest farmers in a village had to contribute to the costs of enclosure. Small farmers suffered because the costs of fencing a small farm were disproportionately greater than for a larger farm, because they lost land in lieu of tithes, and because their compensation for loss of rights on the common was not adequate recompense. Hoskins considered that the 'small peasant' at Wigston Magna finished up with 'a smaller piece of land than he had before (about one acre in six had been deducted in lieu of tithes) and a larger demand for money than he had ever seen before in his life, the cost of doing something that he had not wanted done'.

Thus it is possible to make a strong case for a considerable reduction in the numbers of a distinctive class in the countryside, and to link their reduction to the process of enclosure, or more specifically to the removal of common rights. These commoners could be called 'peasants' because they retained a degree of independence; they were not dependent on working for others as their sole source of livelihood. Yet despite increases in farm sizes and the undermining of the commoners there were still plenty of family farms in nineteenth-century England: that is farms employing no outside labour. The earliest information we have on employment on a national scale is the census for 1831, where, for England as a whole, there were recorded 236,343 occupiers of land in agriculture of whom 36 per cent were not employing labourers. There were wide varieties in the geography of this percentage as Table 4.14 and Figure 4.4(b) show: in Derbyshire, Rutland, Westmorland and West Yorkshire, over 50 per cent of farmers employed no labour, whereas in Buckinghamshire over 80 per cent were employing labourers. 'Family farmer' and 'peasant' are not necessarily synonymous, but it has been argued that a substantial number of these farmers were more concerned with family needs and neighbourhood obligations than with profits from trade in agricultural products.

Nineteenth-century rural society was, however, characterised by a large number of agricultural labourers, who had no land and had to work for others: the rural proletariat. The origins of this class has been an important issue ever since Marx's analysis of the social relations of capitalist production and was originally seen as the corollary of the decline of the peasant, since dispossessed peasants were assumed to become labourers. It is a fairly straightforward matter to document the growth of landless labourers in the countryside. From his survey of manors in the mid-sixteenth century Tawney estimated that 12 per cent of families had no land and 38 per cent held under 5 acres. These are probably underestimates since manorial surveys conceal subletting. Cornwall concludes on the basis of subsidy evidence that by the third decade of the sixteenth century some 20–25 per cent of the population showed signs of belonging to landless families. By 1688, according to King, two thirds of the households were landless. By 1851 agricultural labourers, shepherds and farm servants formed some 73 per cent of all those working on the land; if servants are excluded, the figure falls to 66 per cent. Table 4.14 gives an indication of the regional extent of proletarianisation using evidence from the 1831 and 1851 censuses. It has already been pointed out that the 1851 census understates the number of small farmers, so that the differences between the figures for 1831 and 1851 exaggerate changes between the two dates. What is clearly evident from the table is the regional diversity in employment characteristics. The extent of proletarianisation is indicated by the columns showing the number of day

Table 4.14 *Employment in English agriculture in 1831 and 1851*

	% of occupiers employing labour		Day labourers and servants per occupier not employing labour		Servants as % of servants and labourers	
	1831	1851	1831	1851	1831	1851
Bedfordshire	74	89	24	79	18	5
Berkshire	79	90	32	109	20	7
Buckinghamshire	83	88	37	63	15	7
Cambridgeshire	66	67	12	18	21	4
Cheshire	52	50	4	4	47	26
Cornwall	56	56	4	4	38	23
Cumberland	56	51	3	3	54	38
Derbyshire	44	38	2	2	46	25
Devon	74	71	11	10	35	22
Dorset	70	80	15	31	19	6
Durham	59	54	5	5	39	24
Essex	84	89	43	77	24	5
Gloucestershire	66	74	11	23	25	8
Hampshire	69	81	20	42	24	6
Herefordshire	60	66	7	12	29	12
Hertfordshire	79	90	37	91	23	7
Huntingdonshire	68	80	15	33	17	7
Kent	67	80	17	34	30	10
Lancashire	41	38	2	3	60	20
Leicestershire	55	67	5	9	28	14
Lincolnshire	53	54	5	6	30	14
Middlesex	68	84	23	55	27	6
Monmouthshire	59	53	4	4	42	29
Norfolk	66	75	14	22	21	7
Northamptonshire	73	86	16	47	20	8
Northumberland	65	66	8	12	38	18
Nottinghamshire	52	58	5	7	30	17
Oxfordshire	82	87	35	55	20	6
Rutland	50	57	5	6	24	7
Shropshire	64	68	8	11	36	21
Somerset	62	73	8	15	26	11
Staffordshire	51	52	5	6	36	17
Suffolk	80	85	29	44	22	9
Surrey	72	81	23	39	28	7
Sussex	70	81	20	36	31	8
Warwickshire	71	72	14	18	24	11
Westmorland	46	40	2	2	62	42
Wiltshire	73	83	20	52	17	6
Worcestershire	68	70	12	14	26	9
Yorkshire East	66	65	7	7	37	26
Yorkshire North	53	52	3	4	41	23
Yorkshire West	40	38	2	2	45	18
England	64	69	14	27	31	14

labourers and servants per occupier not employing labour, in other words the ratio between 'peasants' (defined as family farmers) and the proletariat. This is mapped in Figure 4.4(a) and shows how closely the geography of proletarianisation was associated with the grain-growing south and east of the country. By 1831 the south and east (with the exception of London) were also the areas where farming dominated the economy as Figure 4.5(a) illustrates. The county grid is rather a crude one, and more subtleties to the pattern would emerge if the data were broken down into smaller areas.

Debate about the growth of the proletariat has focussed on two periods; the sixteenth century and the late eighteenth century. In both cases the original argument was that the proletariat was an institutional creation. Tawney argued that the rural proletariat was created by the forcible eviction and dispersal of whole communities, the enclosure and conversion to pasture of arable commonfields, the division and enclosure of common grazing land, and the raising of entry fines to unreasonable levels. We have already seen that Tawney underestimated the power of the law in supporting tenants against lords, and that he failed to distinguish between eviction enclosures and abandonment enclosures. He was also writing without the benefit of the knowledge of population change we now have, and it is likely that the growth in the number of labourers, and also of vagrants and vagabonds, in the sixteenth century was due more to population growth than to eviction. Even so, one estimate reckons that some 34,000 families were dispossessed by enclosure and engrossment between 1455 and 1637. Similarly the Hammonds, following Marx, argued that the proletariat was augmented by those dispossessed by enclosure. In many cases this was correct, in fact new quantitative evidence has demonstrated it to be the case. It has also been shown that seasonal unemployment increased with enclosure and there is also a very close correlation between the proportion of a county enclosed and *per capita* expenditure on poor relief for the first thirty years of the nineteenth century. But it is also true that the very rapid growth in population was also responsible for swelling the numbers of the proletariat.

The proportion of the labour force employed as servants was falling as the proportion of day labourers rose. Generally speaking, the incidence of servanthood in the early modern period was inversely related to food prices. Thus the proportion of servants (as indicated by the proportion of October marriages) fell during the sixteenth century, reaching a low point in the middle of the seventeenth century. Thereafter the proportion increased, reaching a peak in the 1740s. When the price of food rose the cost of keeping servants also rose, since they received board and lodging from the farmer. Under these conditions farmers preferred to pay for their labour in cash (to day labourers) rather than hire servants. From the mid-eighteenth century, the sustained rise in prices encouraged a move away

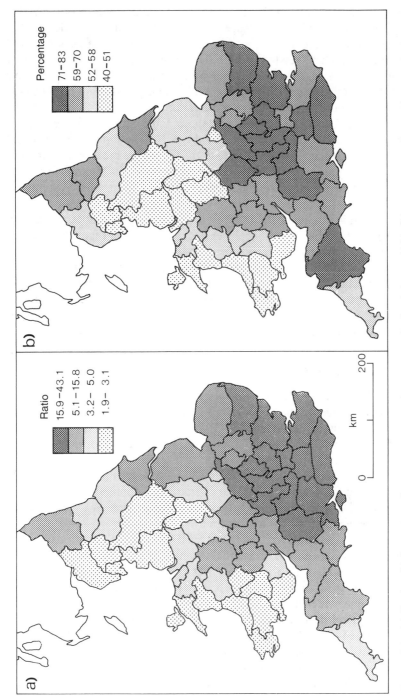

Figure 4.4 (a) Ratio of labourers and servants to occupiers not employing labour, 1831; (b) percentage of occupiers employing labour, 1831. *Source:* Abstract of the population returns of Great Britain, 1831, BPP, 1833, VI.

from living-in servants which continued into the nineteenth century. Aside from the cyclical shift in real wages, the decline of servants reflects more fundamental changes. Farm servants were not suited to large farms and their disappearance is related to both the chronology and the geography of increasing farm size. The greater range of specialist work on large farms suited the short contracts of the labourers, and large farms did not have the provision to house large numbers of servants.

The proportion of the agricultural labour force employed as servants is indicated in the last two columns of Table 4.14. The figures for 1831 and 1851 may not be exactly comparable, so the figures are most reliable in indicating the geography of servanthood, which is shown in Figure 4.5(b). The map shows that servants were associated with the areas where capitalistic farming, defined in terms of Figures 4.4(a) and 4.4(b), was least prevalent, broadly speaking in the north and west of the country. Counties with a high proportion of arable commonfield enclosure shown in Figure 4.3(a) also had a low proportion of servants. Yet, in some areas servants continued to make up a sizeable proportion of the workforce, even in 1851. In East Yorkshire, where the enclosed wolds enjoyed capitalistic high farming, over a quarter of the labour force were servants in 1851. In this area the number of servants was expanding, and although on some farms servants no longer lived in the farmhouse special accommodation was built for them.

Changing social relationships

Changes in the composition of social groups in the countryside were associated with changes in the relationships between the groups. The social relationships discussed in Chapter Two were divided into forces of identification and forces of differentiation. By the nineteenth century the latter were dominant in many parts of England although the former remained of importance. The decline of subsistence farming, the increase in farm sizes and rural proletarianisation, and the dominance of leasehold farms, meant that relationships between individuals were increasingly characterised by explicit or implicit contracts.

An obvious example of such a formal contractual relationship was between landlords and their leasehold tenant farmers. The two groups were not always at odds with one another, yet their interests conflicted, particularly over the proportion of farm profits which should be taken as rent. There is some evidence that when farming was less prosperous, during the so-called agricultural depression of the 1730s for example, landlords in the worst affected areas (those growing grain) would go out of their way to help tenants by allowing them to defer rent payments, and, on occasion, writing off rent arrears. Similar action also occurred in the early 1820s when farm incomes suffered. Landlords were not acting without self-

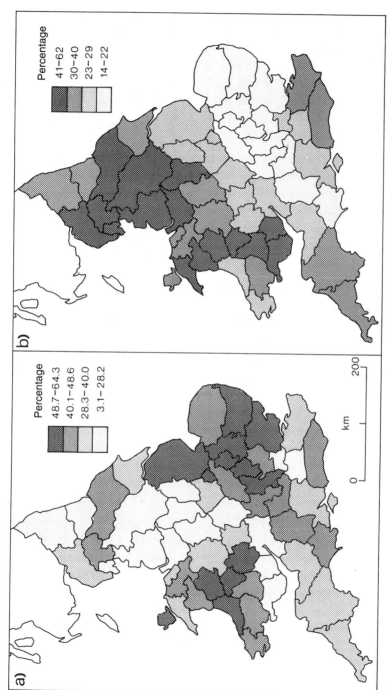

Figure 4.5 (a) Percentage of families employed in agriculture, 1831. *Source:* 1831 Census enumeration abstract, BPP, 1833, VI (b) Servants as a percentage of servants and labourers, 1831. *Source:* Kussmaul (1981), 170–1.

interest, because in the long run, rents depended on tenants' prosperity and profits, and in the short run an untenanted, uncultivated farm would quickly deteriorate.

When times were good, and rising prices enabled farmers to increase their profits, there is evidence that landlords were able to claim an increasing proportion of profits as rent. In the sixteenth century, rents rose faster than prices, and the same phenomenon is apparent again in the eighteenth century. Judging by the evidence of Table 4.6, contemporaries thought landlords' rents were rising more rapidly than farmers' profits. Developments already discussed in this chapter, whereby landlords moved towards leasing their farms at rack rents, contributed to this reallocation of surplus.

The lease could also give landlords more control over the actual farming activities of their tenants. It was in a landlord's interest to encourage good farming practices, and covenants in leases from the eighteenth century stipulated how the land should be farmed, what crops should be grown and what crop rotations followed. Examples include prohibiting the export of manure off the farm, and the stipulation that two grain crops should not be taken in succession on the same piece of ground. However, while some landlords used their leases as a way of promoting innovation by their tenants, it seems that most played safe, encouraging their farmers to stick to well-tried and established practices.

By the nineteenth century, the most contentious issue between landlords and tenants was the issue of 'tenant right' or the extent to which tenants were to be compensated for improvements they made to the farm while they were leasing it from a landlord. The convention was that landlords provided fixed capital and tenants the working capital, but the distinction was sometimes hard to draw. Improvements such as drainage, fencing or the erection of new buildings could be paid for by the tenant but their benefits might well outlast the duration of their tenancy, and by law a tenant's fixtures became the property of the landlord at the end of the tenancy. The problem was exacerbated in the 1840s when off-farm inputs of feedstuffs and fertilisers became common, supplementing supplies produced on the farm, for they made a new contribution to the long-term improvement of farm land. Until the last quarter of the nineteenth century these relationships were governed by custom: it was not until the passing of the Agricultural Holdings Act in 1875 that the tenant received the legal right to compensation for improvements.

Another source of friction between landlords and other social groups, which reflects the changing nature of social relations in the countryside, was over the 'game laws'. The Game Act of 1671 made the hunting of game (particularly hares, pheasants and partridges) the exclusive privilege of the landed gentry, since only those with a freehold worth £100 a year or a leasehold worth £150 a year could hunt game, even if it was on their own land.

Many more acts followed, but the principle remained the same: game was the exclusive property of a specific social class, and for this reason the game laws have been described as 'taking food from the poor to give sport to the rich'. Penalties were also severe for killing animals not regarded as game, such as deer and rabbits. From 1723, for example, the penalty for the armed hunting of deer at night in disguise was death.

Despite these forces for change, the old relationship of paternalism and deference was still very evident well into the nineteenth century, especially in those communities with a resident landlord who dominated landholding and employment in the village. Yet by the eighteenth century, it has been argued that paternalism had become 'as much theatre and gesture as effective responsibility'. Thompson believes that the gentry gradually retreated from paternalism, took less and less interest in controlling the life of the rural labour force, and withdrew from public view and face-to-face contact, preferring to commodify concessions and perquisites wherever possible. Examples of this process include the decline of servanthood, and the attack on the perquisite of gleaning. These changes have been described as part of a 'custom to crime transition' along with the criminalisation of other customary practices and former common rights such as gathering wood for fuel. It can also be argued that this growing intolerance by the gentry of the customary economic rights of the poor extended to other aspects of their life, including recreation, festivals and sporting events. With the decline of paternalism came the decline of deference, and a growing independence for the rural labouring poor, who were less reliant on a particular employer.

Conditions of employment varied considerably across the country: it is obvious that conditions of work and relations between employer and employee would be different on, say, a small farm in the north-west compared with a large capitalistic grain farm in the south-east. According to a mid-nineteenth-century commentator on Norfolk farming: 'After the elaborate and, we may almost say, paternal methods pursued in the north, the Norfolk system of labour is not very attractive. There is no such thing as a yearly labourer, no boarding paid for by the farmer, and, in short, no connection between master and man except work on the one hand and payment on the other.'

Aside from tracing the more general forces influencing the situation of the rural workforce, it is worth attempting to measure living standards more directly. Wage rates on an hourly, daily or piece-work basis can be divided by the price of foodstuffs (just wheat in Figure 3.4), to give an idea of the standard of living. However, this conceals changes in the extent of employment (or unemployment) for the average labourer which might well have changed over time. It ignores sources of income other than wages including common rights to grazing or fuel, the produce from cottage gardens, and payments received from the poor law or local charities. With the employment of servants, it is necessary to give a monetary value to the benefits of living on the farm such as board and lodging. Even labourers received additional payments from farmers, such as a bushel of wheat, keep for a pig, or the right to run one or two animals with those of his employer.

Over time, the simple index of real wages in Figure 3.4 shows a fall in the sixteenth century as the growth in agricultural prices outstripped the growth in wages. Given population growth during the period and contemporary comments about unemployed people, there is no reason to suppose this apparent fall in real wages was mitigated by increased employment. On the other hand many employees were employed as servants in this period, and even those employed as labourers often received additional benefits from their employers as well as customary rights. By the early seventeenth century the fall in real wages as measured in Figure 3.4 was coming to an end and from then real wages rose, to peak in the 1740s, a decade described as a 'golden age' for the English labourer. Over the next century however, real wages once again took a downward path, only showing a sustained tendency to rise from the 1820s. It is likely that the fall in real wages was exacerbated by the removal of common rights, decline of servanthood and reduction in female employment. Moreover, when the position of the farm labourer is compared with other trades, it is clear that *relatively* they were becoming worse off.

There was a change in the regional pattern of wage rates in the eighteenth century. In the early part of the century, as had been the case for several centuries, wage rates were higher in the south and east of the country, reflecting the prosperity of grain farming and the impact of the London market. By the early nineteenth century, however, the situation had changed, and wages were higher in the northern counties (north of a line from the Wash to the Dee, including Staffordshire). For example, in the 1760s Buckinghamshire wages were over 20 per cent above farm wages in Lancashire; from the 1790s onwards, wages in Lancashire remained between a third and a quarter greater than those in Buckinghamshire. This change reflects the effects of industrial wages in bidding up wage rates and the increased demand for agricultural products in the growing northern towns. But it might also reflect a labour productivity difference if northern workers were working harder, perhaps because they were better fed.

Discussion of the fate of the rural workforce would not be complete without some mention of contemporary efforts at what today we call 'social security' and what in the past was referred to as 'poor relief'. Although the parish had been responsible for collecting money to relieve its poor from 1572, in the sixteenth and early seventeenth centuries most help for the poor was provided by charity. But by 1700 most poor relief was administered through the poor law. The poor law act of 1601 established the principle that the parish should provide money levied from a tax on property to support the old, sick and disabled in their own homes, and should find work for the destitute but able-bodied, and apprenticeships for children. The actual practice varied considerably from parish to parish but it seems that most relief took the form of money payments and that schemes to provide work were mostly failures. From 1662 the administration of relief was complicated by the Act of Settlement. This stipulated that a parish was only responsible for the poor who had a legal 'settlement' in its boundaries. Those moving to another parish could get a 'settlement certificate' which stated that their existing parish would be responsible for their poor relief should it prove necessary. The Act of Settlement therefore acted as a constraint to migration, but its impact was felt most severely not by able-bodied males, but by those on the move who were likely to become a charge to the poor rate – especially single women with children.

As the eighteenth century progressed, poor law expenditure continued to increase, and the system was put under extreme pressure in the 1790s when rapid rural population growth coincided with bad harvests and high prices. In 1795, the magistrates at Speenhamland near Newbury in Berkshire introduced a modification to the system of relief which was copied in other parishes in the south and east of England, and became known as the Speenhamland system. Wages were supplemented from the poor rates on a flexible scale that varied with the price of bread. Speenhamland might be interpreted as a late manifestation of paternalism, protecting the rural labourer from the effects of increased wage dependency, but the consequences of the system served only to make matters worse. Those not employing labour now subsidised those that did, farmers had no incentive to raise wages, labourers had no incentive to raise their productivity and were now tied to the poor law, whether they liked it or not. It may also have been the case that the Speenhamland system increased pauperisation by encouraging farmers to lower wages.

The Speenhamland system was swept away, along with the settlement legislation and the rest of the old poor law, by the new poor law of 1834. Relief for the poor was to be provided in the workhouse, and outdoor relief was abolished. The workhouse was designed to be inhospitable and unattractive; for example, family members would be split up if they entered the workhouse. This reflects a powerful ideological change, for whereas the Speenhamland system reflects a paternalistic attitude, the new poor law reflects the imperatives of a market economy with its assumption that poverty is the sole responsibility of the individual.

There can be no doubt that the poor law was central to the life of rural society. Almost all were affected by it, whether as taxpayers or as recipients of relief, and its administration reinforced the social bonds within the village. Aside from payments to those in need, the operation of the poor

law could fix prices, set wages, regulate apprenticeship and control settlement in the village by outsiders. The poor law also had an influence on
other aspects of the rural economy. In comparison with other European
countries, the English poor law was comprehensive in coverage, generous
in benefits and fairly uniform from place to place. It provided a safety net
and a degree of social security not found elsewhere. It can therefore be
argued that the poor law facilitated the movement of people from the land.
In other countries (for example in Ireland until 1838) with no equivalent
poor law, excessive fragmentation of land ownership occurred since people
clung to the land for security. Thus the English poor law encouraged the
growth of wage labour, the decline of servants and the growth in farm sizes.

Another facet of agricultural employment in the eighteenth century was
the changing nature of women's work. In the sixteenth and seventeenth
centuries women were engaged in much the same tasks as men, with the
exception of some of the heaviest farm work like ploughing. Women were
usually solely responsible for looking after the dairy, the pigs and poultry,
the vegetable garden and the orchard. In addition, they frequently were
responsible for marketing produce, and for supervising female servants. As
the eighteenth century progressed, their employment patterns changed and
a more explicit sexual division of labour developed. Increasing farm sizes
meant that there was less necessity for women to labour on the land; the
decline in the incidence of service meant they had fewer servants to admin-
ister; and the growth of middlemen meant that they had less need to go to
market. At one extreme, farmers' wives were getting richer, and no longer
had to suffer the indignities of manual work; at the other, women who
needed to labour on the land found less work available. For example, the
change to harvesting wheat with a sickle instead of a scythe replaced female
with male labour and tended to de-skill women's work. Thus by the end of
the eighteenth century women's labour became marginalised, and confined
to certain tasks: weeding, hoeing, stone picking, dibbling, setting peas and
beans, and leading horses. Harvest work in particular increasingly became
dominated by male labour from mid-century onwards. In areas where
common rights were removed by enclosure, women also suffered since they
were often the main beneficiaries of such rights. However, there were exceptions to this general trend. During the Napoleonic Wars, for example,
women had more opportunities for working on the land since, in some
areas, male labour was depleted by demand from the army and navy.

Rural unrest

Throughout the period from the sixteenth to the mid-nineteenth century
there were periodic outbreaks of violence and unrest in the countryside,
which shed light on the development of social relations and on the work-

ings of the rural economy more generally. Riots were usually associated with a sudden rise in the price of food, but to see them simply as an unreasoned response to hunger serves both to trivialise them and to miss the opportunity of using the evidence they provide to gain a deeper insight into rural society. Disturbances were related to many factors, but it is possible to classify rural unrest into three broad categories: food riots, enclosure riots, and disturbances in the nineteenth century by agricultural labourers, although it should be borne in mind that this classification is rough and ready and some disturbances might fall into all three categories.

Food riots were usually in areas characterised by relatively large numbers of people dependent on a cash income and not on subsistence farming. Agricultural labourers had options for finding food at times of shortage, either through the generosity of their employers or by pilfering, but industrial employees did not. Thus rioting took place in years of dearth and high prices, in locations where a high proportion of the population were dependent on cash purchases of food. Riots took place in 1586, 1594–7, 1622 and 1629–31; in the last period the effects of dearth were exacerbated by interruptions to the cloth trade. The situation improved throughout the remainder of the seventeenth century although problems began to develop as parts of the country became increasingly specialised in industrial production, for example in Cornwall and the north-east of England, which meant that people became more and more reliant on imported grain. The development of private marketing and the abolition of export controls meant that it was much easier for grain to be sent abroad while people went hungry at home. Thus many food riots, especially in the sixteenth and seventeenth centuries, took the form of a mob preventing the movement of grain out of a region and usually destined for London or the export market. They were often located at a crucial node on the transport network. These riots suggest that regional specialisation was causing stress on the distribution of food in years of dearth, and that the inter-regional trade in grain was not yet sufficiently developed to prevent local shortages.

By the mid-eighteenth century, the character of such riots began to change. They were still directed against the movement of grain, but now attempts were made by the crowd to force grain onto the market, and to set a fair price. This happened in 1756–7 when there was widespread rioting, with attacks on middlemen and corn dealers. Similar riots took place in 1766 in the textile-producing areas of the south-west, but during the 1790s price-fixing riots occurred in many areas of the country, especially in 1794–6 and 1799–1801. The nature of food riots from the mid-eighteenth century reflects the changing nature of the marketing of food, and the role of the small consumer in the marketing process, for, as we have seen, the ideology of regulation to set fair prices and to protect the small consumer was being abandoned. The form riots took illustrates an attempt to restore traditional customs and norms.

Riots against enclosure were often coupled with anti-lord and anti-aristocratic attitudes. Kett's rebellion of 1549 included a demand that lords keep their beasts off the common, and the Oxfordshire rebellion of 1596 included a demand for illegal enclosures to be destroyed. The so-called Midland Revolt in May and June 1607, in Warwickshire, Leicestershire and Northamptonshire, was a co-ordinated revolt against enclosures, and saw an armed gentry force confront a thousand rebels encamped at Newton in Northamptonshire who were preparing to destroy enclosures. The rioters saw a clear link between enclosure for pasture and high food prices, and looked to the government for redress. In 1650, a group of commoners in the Isle of Axholme invaded the new settlement of Sandtoft, newly built on lands drained by the Dutchman Vermuyden. They smashed fences, devastated crops and seized cattle. The following year they destroyed eighty-two houses, a mill, barns, implements and crops. There was relatively little riotous opposition to parliamentary enclosure, perhaps because the process was manifestly legal, and because there was a formal opportunity during the passage of an enclosure act for objection with a petition against the bill.

The last major outbreak of widespread rural violence was the Captain Swing riots of 1830–1, by agricultural labourers. Unrest was precipitated by a sudden fall in real wages, and began in Kent with the destruction of barns. The rioters also destroyed threshing machines which they saw taking winter threshing work away from them. Farmers even colluded in this, encouraging the rioters to demand lower rates, taxes and tithes on their behalf. Attacks were also made on those administering the poor law, the overseers. Riots spread from the south-east to Hampshire and the West Country, where arson, larceny and burglary were added to machine-breaking. Threatening letters were also sent, signed by 'Captain Swing'. The Swing riots have been studied in considerable detail. On a general scale the pattern of riots is easily explicable in that they were confined to the arable farming areas of the south and east (excluding the immediate environs of London) where rural proletarianisation was most advanced and real wages were at their lowest. On a more local scale, there is a variety of explanations as to why riots took place in one village rather than another. What is of most interest, however, is the *form* that disturbances took, and what that reveals about rioters' attitudes to social change.

Riots were disciplined, rarely violent, and hardly ever did the rioters propose a radically new social order. Rioters usually demanded that custom and tradition be upheld, and therefore felt that their actions were legitimate. Common rights, perquisites, face-to-face marketing and other customary practices are part of what E.P. Thompson described as a 'moral economy'. Thompson used the phrase to explain the form of riotous action by the eighteenth-century 'mob', whose actions were 'grounded on a consistent traditional view of social norms and obligations, of the proper economic functions of several parties within the community, which, taken together, can be said to constitute the moral economy of the poor'. Thus riots had a legitimising notion to them; rioters clearly felt their actions were morally correct. Collective action was taken to maintain and protect customary practices, which were reflections of a paternalist society. With the decline of subsistence farming, and the increase in dependence by the poor on the market, it is not surprising that much hostility was directed against perceived abuses of the paternalist marketing system. Grain was brought to the market that was being hoarded by farmers or merchants, or being taken for export, fair prices were set for that grain in the market and it was sold to the poor. Although it has come in for much criticism, Thompson's notion of the moral economy has found widespread support amongst historians and some of its arguments have been extended away from food riots and marketing towards a more general model of the pre-industrial economy.

Conclusion

This chapter has shown that the institutional framework within which farming was carried out changed considerably between the sixteenth and the nineteenth centuries. Underlying most changes in the rural economy, including those considered in Chapter Three, was the increased commercialisation reflected in the development of the market. The rise in importance of middlemen and the growth of private sales by sample outside the formal market place indicate a fundamental change of scale in marketing. Although long-distance trading occurred in the sixteenth century, it was not predominant, and most marketing was confined to local market areas. The key development by the nineteenth century was the integration of these into a national market.

Growing commercial pressures had a powerful influence on the changes in husbandry discussed in Chapter Three, but they also influenced the other institutional changes discussed in this chapter. The gradual elimination of customary tenures and their replacement by leasehold farms was prompted by increasing opportunities for landlords to make money. The disappearance of common rights was also a protracted process, but essentially reflects the replacement of a subsistence-orientated rural economy with one firmly linked to the market. The establishment of private property rights does not seem to have been brought about primarily by illegal coercion as some have argued, but by a variety of means depending on the interplay between national economic trends and the details of local customary practice. In some instances, as in the fenlands for example, force and coercion were involved, but this seems to be exceptional.

The pace of change in tenures and property rights varied in different parts of the country, and in the absence of any national statistics it is difficult to generalise a specific chronology. The best estimates are those of the pace of enclosure, and although enclosure embraced many processes, the rate of enclosure is probably a good reflection of the rate of conversion of common to private property. The extent of commonfields in 1500 is impossible to measure, but despite considerable enclosure in the seventeenth century some 4.5 million acres of arable commonfield were enclosed by act of parliament, when the arable area in 1700 was only 9 million acres.

While the close association between enclosure and property rights is clear enough, the relationship between enclosure and landholding change is much weaker than was once thought. Enclosing and engrossing were two separate processes, and the build-up of large estates owed more to economic pressure than to enclosure. Documented increases in average farm size during the eighteenth century show clearly that the growth applied equally to open-field farms as to enclosed farms.

The extent of the decline of the 'small farmer' depends on definition, as Table 4.13 indicates, for there were still a great many 'farms' in the smallest size bands, although most farms under 20 acres would not have provided a living for a family. On the other hand, the majority of the farmland of England lay in large farms; in 1870 some 50 per cent in farms above 100 acres, and almost 80 per cent in farms over 50 acres. Thus in economic terms the significance of the small farmer may not have been very great, but in human terms the small farmer remained an important phenomenon. The decline of the 'peasant' is more problematic, and again depends on definition. If, however, we adopt a fairly orthodox definition, and define the peasantry as a class who were independent of capitalist production, in other words who were able to live without being full-time labourers, then the peasantry was in decline from the sixteenth century, but was dealt the fatal blow by parliamentary enclosure and the elimination of common rights.

Although over 30 per cent of 'farms' in 1850 employed no labour, there were over twenty-five labourers for every family farmer employing no labour, so the dominant class labouring on the land was the agricultural proletariat. The growth of the proletariat is clearly related to the demise of the 'peasant' as just defined, the growth in farm sizes, and the decline of agricultural servants. The agricultural proletariat was not the institutional creation that Marx claimed it was, since enclosure was but one contributor to its growth (compare the distribution of enclosure in Figure 4.3(a) with the distribution of labourers in Figure 4.4(a)). But nor was the proletariat simply the result of population growth and increased opportunities for employment on the land. As we saw in Chapter Three, the proportion of the workforce engaged in agriculture was falling, and an increasing proportion of the agricultural workers were day labourers.

5

The agricultural revolution reconsidered

This chapter returns to the theme of Chapter One. It re-emphasises the case for an 'agricultural revolution', or rather an 'agrarian revolution' consisting of two related transformations; first a transformation in output and productivity, and second, a transformation in the institutional framework of farming. The chapter concludes with an exploration of some ideas about the driving forces behind these agrarian changes; first in terms of responses to market prices, and second through the social relations of production embodied in agrarian capitalism. Before that, however, the first part of the chapter returns to the question of regional variety in farming discussed in Chapter Two, following a brief comparison of farming in the mid-nineteenth century with farming in the early sixteenth century.

High farming

The apogee of the conventional 'agricultural revolution' in 1850 was 'high farming'. 'High' was used as an adverb meaning excellent, and while many contemporary farmers regarded high farming as excellent, its meaning came to be associated with an intensive system of farming with high inputs and high outputs. The basis of high farming was mixed farming embodying the principles of the Norfolk four-course; but by 1850 the rotation had been extended and production intensified. In many light-soil areas, high farming was intensive mixed farming, with a corn crop taken two years in every four or five, and the remainder of the arable under fodder. By the 1850s, catch crops had been introduced on the short fallows between the major crops, so that fodder crops now included swedes, turnips, rape, vetches, kale, mangolds, rye grass, clover, cabbages, sainfoin and kohl-rabi. In addition to this intensive cultivation, other inputs included imported feedstuffs such as oil-cake, and imported fertilisers such as guano. By 1850, new artificial fertilisers including superphosphate were just coming onto the market, but the main input was fodder, whether grown on the farm or bought in. As

'Alderman' Mechi, one of the more eccentric propagandists for high farming, put it, 'the more meat you produce the more manure you make and consequently the more corn per acre you will grow on the arable portion'. High farming was not ubiquitous, and in some areas, especially on the heavy claylands, wheat, bean and fallow rotations continued much as they had for centuries. The key to increased productivity on clay soils was drainage. The majority of effective drainage took place after 1850, but there was much activity and experimentation before this, often accompanied by sub-soiling to break the plough pan. Deep drainage and deep ploughing were accompanied by greater stocking densities, the buying in of fodder, and the greater production of manure.

The problem with high farming, at least as it was characterised by those contemporaries who espoused its virtues with such vigour, was that maximum production was to be achieved at almost any cost. It is likely that high output per acre was not matched by a high return on capital invested, although it is difficult to calculate a cost-benefit of high farming as a whole. It is also ironic that such an intensive farming system was at its zenith as the corn laws were repealed, exposing English farmers to the potential of competition from the extensive farming systems in North America. The main brunt of this competition was not felt until the last quarter of the nineteenth century, but, even so, whereas English farming was about 90 per cent self-sufficient in 1850, by 1875 it was about 75 per cent self-sufficient. Thus high farming was not a sustainable agricultural system in the context of the world-economy, nor, incidentally, was it sustainable in the sense that an increasing proportion of its inputs were imported from abroad.

The contrasts with farming in 1500 are clear enough. Farming systems at this time were sustainable, albeit at much lower levels of production, saw little capital investment, much lower levels of production intensity, and much less emphasis on fodder crops and livestock production. Yet the sixteenth-century farmer would have seen much that was familiar had he been transported forward to 1850. Most farming operations – ploughing, sowing, weeding and harvesting for example – were carried out in much the same way on the majority of farms, since mechanisation had not made great headway by the mid-nineteenth century. Similarly, only a few farms employed steam power, and only on a minority of farms was the impact of the railway felt directly. In contrast with the present day, the most striking similarity between the sixteenth and the nineteenth centuries was the inability of farmers to control pests and diseases, and the rudimentary nature of agricultural science, which, by 1850, was doing little to help the ordinary farmer.

Despite a fivefold increase in agricultural output, it is not changes in production that make the most striking contrast between farming in 1500 and farming in 1850. In 1500, more than half of the arable land of England lay in commonfields; by 1850, common property rights on arable land had almost disappeared. Around 75 to 80 per cent of the land of England was controlled by landlords who leased their land on short-term tenancies to tenant farmers. The vast majority of those working on the land – the agricultural labourers – had no land of their own and employment in a single occupation had replaced the multiplicity of occupations that individuals often had in the sixteenth century. There are exceptions to these generalisations about the rural class structure in the mid-nineteenth century; there was still a sizeable body of family farmers for example, but the new class relations dominated agricultural production.

Changes in property rights, tenures and class composition reflected a fundamental change in attitudes to agricultural production. Many farmers in the nineteenth century were producing for consumers in distant markets, separated from them by a growing army of middlemen, dealers and processors of food; production was for exchange in the market, rather than for use or subsistence. In the sixteenth century, the majority of farmers were producing at subsistence levels; although they might have been engaging with the market, for the majority of farmers their produce was confined within their local trading network and the market was not the major determinant of production. In the nineteenth century, most of the farmland of England was farmed by farmers for whom farming was a business activity as much as a way of life. Once this is the case, and agriculture becomes a means of making money, it follows that adaption and innovation become the norm (albeit at a slower rate than for other sectors of the economy).

Agricultural regions

Better farming practices made farmers less dependent on the inherent characteristics of the soils on their farm, and this was one of many reasons why we might expect there to be less regional differentiation and greater uniformity in nineteenth-century farming. On the other hand, increased commercialisation and economic integration encouraged specialisation at both a farm and a regional level, which would encourage diversity. The most spectacular cropping changes were on land that had been reclaimed, especially the fens and the lowland heaths. The battle to reclaim the fens was not won in 1850, but the change to the landscape was nevertheless dramatic. Lightlands and heathlands saw the cropping changes that are most characteristic of the 'agricultural revolution': the introduction of fodder crops and rotations based on the principles of the Norfolk four-course rotation. These lightlands were areas of relatively poor soils, which had hitherto been under pasture (compare the lightland in Figure 2.10 with the good-quality land in Figure 2.11). They also tended to have lightly regulated, flexible field systems, and were able to adapt relatively easily to innovation. Such innovation was not inevitable, of course, and within these lightland areas, the most favoured locations were those with good transport links for taking grain to market (both at home and overseas).

An example is provided by a comparison of the wolds of East Yorkshire with the Wessex downlands. Both areas were very similar in terms of soil, topography and climate, and until the 1770s enjoyed a similar sheep-corn husbandry system. But over the succeeding years, wolds farming developed into an intensive, high-input system, while the downlands remained as a low-intensity sheep-corn system. The difference can be explained by differences in market demand. In the north, new urban centres provided a strong demand for mutton which led to an emphasis on fodder crops and the high feeding of improved sheep such as the Lincoln and the Leicester. Improved down breeds appeared in Wessex, but they were developed for their role in folding rather than for their mutton, and as a consequence fewer fodder crops were grown.

Paradoxically, areas of good-quality land, especially in the Midlands, were less adaptable. Although the land was of good quality, it was rather heavy for the introduction of root crops, and many communities were hidebound by the rigidities of a commonfield system. When this system was removed by enclosure, rather than introducing cropping innovations many of these areas went over to permanent pasture, especially during the first phase of parliamentary enclosure from 1750 to 1790.

This transformation was commented upon by Caird as follows:

In former times the strong clay lands were looked upon as the true wheat soils of the country. They paid the highest rent, the heaviest tithe, and employed the greatest number of labourers. But modern improvements have entirely changed their position. The extension of green crops, and the feeding of stock, have so raised the productive quality of the light lands, that they now produce corn at less cost than the clays, with the further important advantage, that the stock maintained on them yields a large profit besides.

New regional patterns were also evident in other aspects of the rural economy. Agriculture no longer dominated the economy of the entire country, but was relatively much more important in the south and east (Figure 4.5(a)). It was in these areas that agrarian capitalism was most firmly rooted (Figure 4.4) although by 1850 the geography of field systems was uniform with the universal establishment of private property rights and enclosed farms.

In comparison with work on the early modern period, little attention has been paid to the geography of nineteenth-century farming. Although farming regions have been mapped, they are simply amalgamations of the regions produced by separate authors for the 'prize essays' on the agriculture of each county in England published in the *Journal of the Royal*

Agricultural Society of England between 1844 and 1869. More specific farming regions, constructed on a more consistent basis, have yet to be produced.

The regional expression of relationships within the early modern rural economy discussed in Chapter Two loses much of its relevance in the mid-nineteenth century. Agriculture was less dependent on soil type than it had been in the sixteenth century; industry and agriculture were separate occupations; and the disappearance of commonfields and establishment of new class relations meant that individualism versus collectivism was no longer an important issue. The one sixteenth-century theme that retains its relevance for the nineteenth century is that of social control in the local community. The local impact of landownership could influence religion, politics, employment and settlement, and was responsible for considerable parish to parish variations in the mid-nineteenth century. Many parishes were neither 'open' nor 'close' but there were sufficient in both categories to influence the economic and cultural geography of the countryside on a parish scale.

Output and land productivity change

The comparison of the agrarian economies of the sixteenth and the nineteenth centuries prompts the question as to which of the many differences might be considered 'revolutionary'. It could be argued that changes in the look of the landscape, the abolition of commonfields or the emergence of a new class structure amounted to an 'agricultural revolution'. However, the case has already been made in Chapters One, Three and Four for an 'agricultural revolution' based on a transformation in output and productivity, and a transformation in the institutional framework of agricultural production. Although there were some productivity improvements in the seventeenth century, especially regarding livestock, they cannot compare with the magnitude of changes in the eighteenth century. Similarly, although there is evidence of improvements in farming methods from the late sixteenth century, it was not until after 1750 that high-yielding fodder crops were grown on a substantial scale, enabling intensification of production through a reduction in fallow and a massive increase in the supply of nitrogen to farmland.

Thus the arguments for an 'agricultural revolution' commencing in the sixteenth century fail to carry conviction. There is some justification in the claim that breaking the distinction between pasture and arable is revolutionary, or at least is a change of potentially revolutionary significance, although the evidence on which the claim is based is open to varying interpretations. The ploughing up of pasture land can also be interpreted as a desperate attempt by farmers to cash in on reserves of nitrogen to produce as much grain as possible in the face of overwhelming demand. Putting land back under a temporary ley would be much more difficult, and it was not until clover and other grass seeds became more widely available in the eighteenth century that true convertible husbandry could take place. For all his volume of footnotes, Kerridge's arguments are not persuasive, and the moderate rise in yields from the mid-sixteenth century was most likely the consequence of increased labour inputs, and labour productivity was probably falling from the mid-sixteenth to the mid-seventeenth centuries. Coupled with evidence of widespread reclamation and the halt to population growth in the mid-seventeenth century this period is more suggestive of a Malthusian check than an agricultural triumph.

Nor is there any evidence to suggest that changes in the century after the Restoration were of more significance than those that were to follow. Admittedly English agriculture had achieved an export surplus by 1750, and output was growing at a faster rate than was population. It was also the case that in some areas crop yields may have been rising (although they were still within medieval norms until the eighteenth century); but rises in yields were not yet associated with the introduction of new crops. In fact the changes of most significance were concerned with livestock husbandry: in Norfolk, there was a remarkable doubling of livestock densities in the seventeenth century reflecting improvements to fodder supplies which are less conspicuous than the innovation of turnips and clover. On a national scale the evidence from prices suggests an improvement in the yield of both wool and mutton during the first half of the eighteenth century. These developments were taking place at the same time as a steady improvement in labour productivity after 1670, suggesting an increase in the overall efficiency of agricultural production. But despite these developments, it was not until the century after 1750 that the dramatic and unprecedented improvements in output, land productivity and labour productivity, associated with equally dramatic and unprecedented changes in husbandry, were under way on a broad front.

How can we account for the timing of this change? Hitherto, emphasis has been placed on the relationship between population growth and agrarian change, and the close association between the rate of growth of population and the rate of growth in prices has been demonstrated in Figure 3.5. The cessation of English population growth in the mid-seventeenth century has been interpreted as a Malthusian preventive check: agricultural supply could not meet the increase in demand prompted by population growth. Falling real wages encouraged couples to delay marriage, the birth rate fell, and population stopped growing. Figure 3.5 shows how this Malthusian framework breaks down in the late eighteenth century, so while it is an appropriate structure in which to understand agricultural change before that date, it cannot by itself explain why land productivity rose so significantly.

Nor is the alternative hypothesis of the relationship between population growth and agrarian change put forward by Boserup much help in this context. She argues that population growth provides the stimulus for the development of agricultural technology. Yet in its essentials the agricultural technology of the eighteenth century was not new. Most of the components of the technological package embodied in the Norfolk four-course rotation – the use of legumes, a reduction of fallows, manure-intensive husbandry, and the integration and mutual development of arable and pastoral husbandry – had been available to certain English farmers in the middle ages. Turnips were known as a fodder crop from at least the 1630s, yet they were not widespread for another 150 years, and it was another century before they reached the peak of their adoption by farmers in England. Clover appeared a few years after turnips, in the 1650s, but thereafter the chronology of the crop's adoption was similar. It is of course true that the widespread innovation was accompanied by very high rates of population growth in the late eighteenth and early nineteenth centuries, but very little innovation took place in the previous period of sustained population growth from the mid-sixteenth century. In any case Boserup's arguments were never really intended to be applied to a market economy like eighteenth-century England.

As an alternative to a population-resources framework, economic historians have favoured interpretations of agrarian change based on the behaviour of farmers in response to changes in market prices. In simple terms, when farmers' perceptions of trends in prices and costs suggest to them that they will make more profit by changing their farm enterprise, then they will do so, including the adoption of new techniques. Rational responses to price movements underlie Jones' interpretation of agricultural change in the century after 1660. He interprets the innovation of fodder crops as an attempt by farmers to cut costs in the face of falling grain prices, which is a perverse response to prices since output expanded in the face of falling prices. Modern studies suggest that farmers try harder to maintain their incomes than they do to raise them, and we would expect farmers to produce more grain (given that no farmer or group of farmers could raise prices by withholding supplies) if unit costs could be lowered. Of course, it is equally rational for farmers to expand grain output when prices are rising; so prices can be persuaded to provide an 'explanation' of changes in production whatever the relationship between the two might be.

The production of new evidence during the last thirty years necessitates some modification of Jones' ideas. The evidence of prices, for example, shows that the swing towards livestock is not uniform over the various grains and livestock products. Table 5.1 and Figure 5.1 show that the trends in price ratios were quite complicated. While wheat prices fell

Table 5.1 *Livestock and crop price relatives, 1550–1750 (1550–1750=100)*

	Cattle/ Wheat	Beef/ Wheat	Sheep/ Wheat	Mutton/ Wheat	Wool/ Wheat	Cattle/ Barley	Beef/ Barley	Sheep/ Barley	Mutton/ Barley	Wool/ Barley
1550s	83	–	76	–	115	–	–	69	–	103
1560s	85	91	81	–	104	81	–	73	–	92
1570s	83	68	77	–	87	78	–	90	–	99
1580s	89	76	78	–	86	85	–	88	–	95
1590s	68	60	70	–	76	67	–	78	–	83
1600s	89	86	84	101	110	80	94	79	94	102
1610s	90	78	77	88	90	71	79	69	79	80
1620s	97	79	77	91	91	76	88	75	88	88
1630s	75	63	63	70	77	64	71	64	71	78
1640s	83	74	77	78	76	73	77	76	77	74
1650s	–	–	–	–	–	85	100	76	100	106
1660s	96	99	84	118	127	101	121	87	121	130
1670s	99	95	88	106	111	100	111	93	111	116
1680s	106	109	108	124	108	106	120	106	120	104
1690s	106	89	99	100	111	99	111	112	111	123
1700s	115	111	124	106	120	98	94	111	94	106
1710s	104	113	103	88	87	138	107	126	107	105
1720s	122	136	126	98	89	127	92	119	92	83
1730s	157	194	179	121	122	195	121	180	121	121
1740s	155	179	229	112	112	177	111	229	111	110

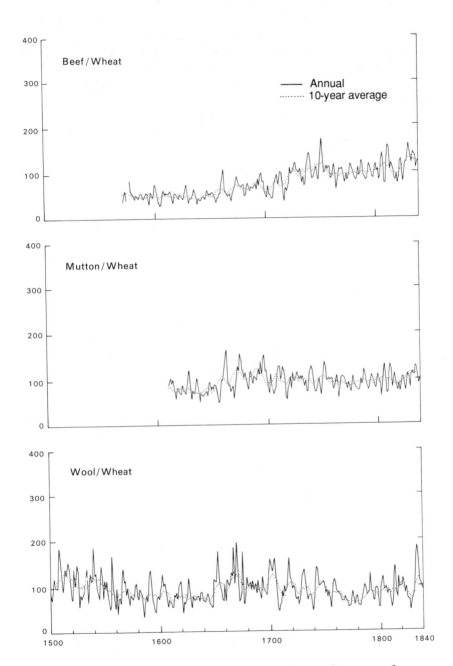

Figure 5.1 Agricultural price relatives, 1500–1840. *Source:* See sources for Table 3.1.

considerably relative to beef, the price of mutton and wool showed little change in relation to barley in the eighty years from 1660. These price signals are complicated further by government subsidies, especially for barley. From 1697, excise duties were waived on malt destined for export, which effectively raised the price to the farmer by 14 per cent. But however farmers interpreted the market it is most unlikely that fodder crops like turnips and clover cut costs through raising yields before 1760. We have seen that although the crops were common their acreages were comparatively small and that turnips were not cultivated in the manner they were to be a century later.

Using prices to explain agricultural change in this context is also difficult because of uncertainties about the extent to which farmers' production decisions were determined by movements in market prices. Discussion of farmers' responses to prices in the early nineteenth century are fairly convincing in showing the link between price movements and production changes. During the Napoleonic Wars, for example, Northumberland farmers adopted a strategy of convertible husbandry to take advantage of high prices for both grain and meat, and a study has shown how developments in sheep breeds corresponded to the changing relative prices of wool, mutton and tallow. The situation is more problematic for the period 1660–1760. Farmers would have been well aware of short-term fluctuations in grain prices but were they aware of longer-term trends? It must be of some significance that no agricultural price series were published (with the exception of Houghton's from 1692 to 1703) until the mid-eighteenth century, and despite strenuous efforts, so few price series have been unearthed from the archives. Demonstrating a necessary link between farming behaviour and price movements is difficult: it is especially difficult when the proportion of farmers basing their decisions on the dictates of the market, rather than on traditional custom and practice, changes over time.

There is perhaps an inevitable tendency to provide explanations of agrarian change that are *a posteriori*, that is they move from effect to cause. Thus, explanations as to why new fodder crops were being grown in the late seventeenth century sometimes seem to assume implicitly that farmers could look into the future and foresee the benefits that the Norfolk four-course rotation was to bring in the nineteenth century. The initial cultivation of clover and turnips may have had more to do with averting the risk of a failure in fodder supplies than with raising land productivity through the complicated mechanisms of the Norfolk four-course. A problem with ley farming or convertible husbandry was the difficulty of establishing a grass ley. The initial attraction of clover may well have been its ability to form a ley more quickly and reliably than hay seeds or other kinds of grass, especially in the drier areas of the country. The late seventeenth century marks the peak of the so-called Little Ice Age, characterised by a fall in average temperatures, and farmers may well have turned to turnips with such alacrity to diversify their fodder portfolio. Evidence from a Norfolk weather diary for 1659–85 shows that drought was extraordinarily prevalent in May and June which would have led to poor hay crops. Turnips could be sown for their green tops as late as August to provide winter fodder and were thus a means of mitigating the effects of a poor hay crop. Farmers seldom leave a record explaining their actions, but in 1681 a churchwarden of Hingham in Norfolk recorded: 'This year began a drought about the middle of March by reason of which we had little or no hay, for that it was sold for a great price before the rain. But in July it pleased God to send rain ... the want of hay was supplied by the growing of turnips.'

Thus it is possible that the full benefits of turnips and clover were unintended consequences of their initial innovation in the late seventeenth century. If this is the case then sophisticated arguments about their role in raising yields as a response to price movements lose some of their appeal.

Whether farmers were primarily motivated by the desire to maximise income or minimise risk, their actions were circumscribed by a host of possible restrictions. But by the mid-eighteenth century, there were enough farmers who were not only willing to respond to the increased demand for food arising from renewed population growth, but also able to meet the demand by making full use of the technology then available. That they were able to do so suggests that changes had taken place in the institutional structures under which farming was carried out.

Agrarian capitalism

Chapter Four discussed changes in tenures, property rights, farm size and social differentiation. Historians see the development of these aspects of the agrarian economy amounting to the establishment of an agrarian capitalism by the mid-nineteenth century. Capitalism is a difficult word because it has been used in so many different ways by different authors, but in the context of English agriculture, it is often taken to mean the tripartite class division of the countryside into the landlord, tenant farmer and labourer. But capitalism involves more than this, including the production process itself, as well as the social relations of production. By the nineteenth century, fairly continuous technical change had become an established feature of English farming. Neither the progress nor the impact of the technology was as dramatic as it was in some parts of industry, but the incorporation of technological change meant that nineteenth-century farming was fundamentally and irrevocably different from farming in the sixteenth century. Nineteenth-century farmers would not expect their children to farm using the methods that they were using. The majority of English farmland was cultivated by farmers who were not subsistence farmers, who were competing with other farmers, and whose decisions on what to produce were largely determined by the market. Average farm sizes were now much larger, and many landlords and farmers were employing 'capital' in that they were investing heavily in the long-term improvement of the land. A prerequisite for such investment is private property, since with common rights the return on such investment would accrue to those holding common rights as well as to the owner of the land.

The social groups involved in farming were now dominated by the three classes already mentioned, although there were still considerable numbers of family farmers in the nineteenth century. Landlords owned 75–80 per cent of the farmland of England by 1850 and were responsible for the majority of capital investment in the land, perhaps six times as much as their tenants. The tenants' business was the management of farming, and the provision of working capital, while agricultural labourers supplied their labour, which was their sole means of survival since they had no land. The relationships between these groups were primarily contractual.

This characterisation of capitalism is a considerable simplification of a complicated reality, but it serves to highlight the distinctive and important features of agricultural production. Explaining the development of agrarian capitalism is partly a matter of explaining how these individual changes came about. The emergence of landlords, tenant farmers and the rural proletariat has been discussed in the previous chapter, as has the move towards commercialised market-orientated farming. For Marx, the mechanism for creating these new social relationships (and also for creating the industrial proletariat) was parliamentary enclosure, although it is clear that he exaggerated the importance of this process. Enclosure, both parliamentary and non-parliamentary, could bring about changes in landownership, tenures, property rights and farm sizes, but, as we have seen, it was not the only means to these ends.

Other ideas about the transition to capitalism in the economy more generally can be grouped into two camps: those who emphasise the importance of the market (the external route) and those who consider changes in the social relations of production to be the driving force behind capitalist development (the internal route). The importance of the market is stressed by many, including those outside a Marxist tradition, as part of a more general thesis that economic development is almost inevitable once barriers to economic activity are removed. Thus, in addition to the development of the market, and especially the reduction in transaction costs, they also see the establishment of private property rights as important to this process.

Those who argue that the relations of production are paramount see conflict between social groups as stimulating capitalist production. The most recent proponent of this view is Brenner, who argues that agrarian capitalism developed in England through the rise of 'landlord–large tenant relations', which amounted to an historical transformation of class structures through class struggle, and could not have been induced by the market nor shaped by demographic swings or price movements. He argues that these new relations were brought into place through landlords forcing out small farmers and creating tenancies and large farms through coercion and class power. Once these new relations were in place, Brenner assumes that technological change and the expansion of agricultural output follows automatically.

The material discussed in Chapter Four lends little support to Brenner's view. There is mounting evidence to show that there was not a coordinated relationship between landlord power, tenure, ownership, farm size and capitalistic farming. Landlords were frequently unable to exercise the power that Brenner attributes to them: customary tenancies and leases could give considerable protection to tenants, whose rights were upheld in the courts. In general, economic differentiation was a process which took place among the tenantry. Moreover landlords, especially in the sixteenth century, showed little interest in developing their estates for capitalist tenant farming, and as a rule they were not very adventurous in promoting innovation in agriculture. The pioneers of new methods in the seventeenth and eighteenth centuries (at least in Norfolk) were not the great landowners but smaller farmers, both tenants and owner-occupiers. Large farms are an obvious requirement for a capitalist agriculture that includes an agricultural proletariat (since large farms employ more labour than small ones), but large farms were not necessarily a prerequisite for higher land productivity. Several studies have now shown that crop yields were independent of farm size from the seventeenth century onwards. Finally, according to Brenner, agrarian capitalism developed in those areas where the control of lordship was strong and peasant property rights were weak. In fact the reverse is the case; the most dramatic advances in output and land productivity came in those areas (such as Norfolk) where lordship was relatively weak.

While the market developments discussed in Chapter Four were the essential accompaniments for the increases in production discussed in Chapter Three, farmers still had to respond to the possibilities that market developments offered. The extent to which farmers were responsive to market prices has already been discussed, but the issue is difficult to resolve because they have not left records of how they made production decisions. Instead, we can gain some insight into attitudes to the market from contemporary published literature, although this is as likely to be expressing an ideal as it is reality. Even so, in the late seventeenth century, agricultural production was seen as an activity in which the individual husbandman worked for himself and his family on his own lands. By the mid-eighteenth century the husbandman had become the farmer, and, instead of 'husbanding' nature, was seen as an entrepreneur, calculating the costs and benefits of alternative courses of action. In 1796 Robertson considered the duties of the farmer to be: 'appointing and superintending labour ... to tell others what to do, how to do, and to see it done... . make observations, to think, to read, to go to markets, to meet with his neighbours to ride through Parish and Country, and get information from all quarters'.

Identifying changing attitudes is important because it acts as a corrective to the view that agricultural development automatically follows from changes in prices and costs, in whatever period or historical context. Mentalities and attitudes change, and while the pressures of commercialisation and the market provided the essential stimulus to agrarian change, farmers' attitudes and expectations also had to change.

Conclusion

Claiming an 'agricultural revolution' involves justifying the criteria on which that revolution is based and establishing the empirical evidence to demonstrate that the criteria were met. The criteria for an 'agricultural revolution' adopted in this book are an unprecedented increase in the output of English agriculture that was associated with an increase in the efficiency of production as measured by land and labour productivity. The transformation of output and land productivity enabled the country to break out of a 'Malthusian trap', allowing the population to exceed the barrier of 5.5 million people for the first time. Rising labour productivity ensured that extra output could be produced with proportionately fewer workers, so making the industrial revolution possible. These increases in output and productivity would have been impossible had England remained a country where the majority of farmers were producing at subsistence levels and local markets remained virtually independent of each other. Farming had to become a business, a money-making enterprise, before farmers would take advantage of market opportunities and produce more.

Statistics of output and productivity show that although change was under way by the mid-seventeenth century, it was not until the century after 1750 that the decisive breakthroughs took place. Evidence of changes in farming practice also show that this was the period when change was most rapid and was making the most significant contribution to output and productivity. Labour productivity was also increasing throughout the eighteenth century, but here the reasons for the increase need further investigation.

It is more difficult to measure the rate of change in the institutional side of farming, in other words the progress of agrarian capitalism. Changes in landholding and property rights were underway before major increases in agricultural output, yet the rate of change in output is remarkably close to the rate at which capitalist production relations were being established. Private property was not essential for innovation or agricultural improvement but it certainly assisted it. Innovation took place on both large and small farms, although the heavy capital investment involved in land reclamation and enclosure required farmers or landlords of substance to carry it out. The key to the relationship between institutional change and farming practice lay more with commercialisation and the market than with the social relations of production. The integration of local markets and a new willingness of farmers to exploit commercial opportunities provided the impetus for innovation and enterprise which led to the agricultural revolution.

Sources for tables

2.1 Implements of husbandry for a seventeenth-century farm. Worlidge (1697), 233–4.

2.2 Harvest yield and income by farm size. Wrigley (1987c), 96 for Bouniatian's formula.

2.3 A functional classifcation of arable field systems. Campbell (1981), 113–17.

2.4 Regional variations in landholding. Tawney (1912), 25 and Bettey (1982), 35.

2.5 Occupation and status labels of those engaged in farming recorded in Norfolk and Suffolk inventories, 1580–1740. Norfolk and Suffolk probate inventories described in Overton (1985).

2.6 Wealth and status of Norfolk and Suffolk farmers leaving probate inventories. Norfolk and Suffolk probate inventories described in Overton (1985).

2.8 Wealth distributions in 1522. Cornwall (1988), 33, 36 and 45.

3.1 Wheat, barley, oat, beef, mutton and wool prices, and agricultural wages, 1500–1849 (10-year averages 1700–49=100). 1500–1750: Bowden (1967b), 815–21; (1985b), 828–31. 1750–70: Mitchell and Deane (1962), 486 (for wheat); Beveridge (1939), 146–7 (for barley at Eton), and 83 (for oats at Winchester). 1771–1840: Gazette prices in Mitchell and Deane (1962), 488–9; Beveridge (1939), 144–7 (for mutton at Eton, 1600–1840), and 209–10 (for beef at Charterhouse, 1713–1830); Rogers (1882), IV, 333; (1887), V, 347–53; (1902), VII, 284 (for beef at Cambridge, 1562–1712); Bowden (1967b), 839–45; (1985b), 843–6 (for wool, 1500–1749); Mitchell and Deane (1962), 494–5 (for wool, 1750–1840). Wages: Bowden (1967b), 864; (1985b), 877–8; Mitchell and Deane (1962), 348–9.

3.5 Estimates of English agricultural output, 1520–1851. (a) Population figures from Wrigley and Schofield (1981); import and export estimates from a variety of sources; (b) Crafts (1983), pp. 177–99; Jackson (1985), pp. 333–51; (c) the combined value of the output of wheat, rye, barley, oats, potatoes, mutton, beef, pork, milk, butter, cheese, wool, tallow and hides based on the estimates in Chartres (1985), 406–502; Gregory King's estimates in Thirsk and Cooper (1972), 782–3 and Holderness (1989), 84–189. Prices are from Clark (1991a), 215–16 and John (1989), 974–1009.

3.6 Estimates of land use in England and Wales, c. 1700 – c. 1850. Whitworth (1771), II, 216; Prince (1989), 30; cf. Grigg (1982), 185; Agricultural returns

for Great Britain for 1871, c.460, BPP 1871, LXIX.
3.7 Land productivity estimates, 1300–1860. (a) Volume output figues from Table 3.5 divided by estimates of land area from Table 3.6; (b) Clark (1991a), 215–16; (c) Lincolnshire, Norfolk and Suffolk: Overton (1991), 302–3; Hertfordshire and Hampshire: Overton and Campbell (1991), 40; Glennie (1991), 273; Kain (1986); (d) Turner (1982); Kain (1986); Craigie (1883).
3.8 Labour productivity estimates, 1520–1871. (a) Wrigley (1985), 700–1 and Clark (1991a), 228–9; (b) population-based output figures in Table 3.5 divided by Wrigley's estimates of the rural agricultural population; (c) volume-based output measures divided by Wrigley's estimates of rural agricultural population; (d) Clark (1991a), 215–16.
3.9 Total factor productivity estimates, 1760–1870. (a) Crafts (1987), 251; (b) McCloskey (1981), 114; (c) Hueckel (1981), 192; (d) Mokyr (1987), 310.
3.12 Crop percentages for selected areas, 1530–1871. 1801 data from Turner (1981a), 296; *c.* 1836 from Kain (1986); 1871 from the Agricultural returns (see Table 3.6). Inventory material, Cornwall: Whetter (1974), 46–7; Hertfordshire: Glennie (1988a), 61; Kent: Chalklin (1965), 78; Lincolnshire: Thirsk (1957), 39, 78, 89, 102, 136, 157, 173, 188, 224–5, 246, 280–1 and 302–3; Norfolk and Suffolk: Overton (1991), 306; Oxfordshire uplands: Havinden (1961), 81; East Worcestershire: Yelling (1969), 27.
3.13 Crop proportions in the 1801 crop return. Calculated from Turner (1981a).
3.14 Crop proportions *c.* 1836 and in 1871. 1830s from Kain (1986); 1871 from Agricultural returns (see Table 3.6).
3.17 Evidence of 'grassland' farming in Norfolk and Suffolk, 1584–1739. Overton (1991), 307.
3.18 Ratios of animal and animal product prices, 1550–1750. Grain and livestock prices calculated from probate inventories in Hertfordshire, Lincolnshire, Norfolk, Suffolk, and Worcestershire. Livestock products as for Table 3.1.
3.19 The impact of the Norfolk four-course rotation. Overton (1991), 296.
3.20 Norfolk: trends in agricultural production, 1250–1854. Campbell and Overton (1993).
3.21 Work rates and labour requirements for hand tools and machines. Collins (1981), 210 and (1969), 460.
3.22 Horses and labourers in English and Welsh agriculture, 1700–1850. Wrigley (1991), 328 and (1988), 37–42; Thompson (1976), 80.
4.1 The English non-agricultural population, 1520–1851. Wrigley (1985), 700–1.
4.2 Places wholly enclosed in Leicestershire before 1750. Beresford (1948), 81.
4.3 Enclosure in County Durham, 1550–1850. Hodgshon (1979), 87–8.
4.4 Enclosure in the South Midlands. Allen (1992), p. 31.
4.5 The chronology of English parliamentary enclosure. Turner (1980), 196–207.
4.6 Costs and benefits of four types of parliamentary enclosure (£). Young (1808), 286–7, taken from *The advantages and disadvantages of enclosing waste lands*, by a Country Gentleman.
4.7 Farming before and after enclosure in 1786 at Canwick, Lincolnshire. Young (1808), 270–1.
4.8 The distribution of landownership in England and Wales, 1436–1873. Mingay (1976), 59.

4.9   The ownership of manors in Norfolk, 1535–65. Swales (1966).
4.10 Landholding in Chippenham and Willingham. Spufford (1974), 166–7.
4.11 Farm sizes in the South Midlands. Allen (1992), 73–4.
4.12 Farm sizes on the Leveson-Gower estates. Wordie (1974), 596.
4.13 The distribution of English farm sizes in 1870. Agricultural returns 1870 c. 223; Census of Great Britain, 1851, II, pt I, 1852.
4.14 Employment in English agriculture in 1831 and 1851. 1831: The method is from Kussmaul (1981), 170–1 and the data from Census Enumeration Abstract, BPP 1833, XXXVI, 1851.
5.1   Livestock and crop price relatives, 1550–1750. See Tables 3.1 and 3.18.

Guide to further reading

Chapter One: The agricultural revolution

The debate on the agricultural revolution is reviewed by Overton (1984b, 1986a and 1989a), Woodward (1971), Walton (1990) and Beckett (1990b). The classic statements in favour of the period 1750–1850 are found in Toynbee (1884) and Prothero (later to become Lord Ernle, 1888 and 1912). They were popularised by Curtler (1909), Gras (1925), Orwin (1949) and Whitlock (1965). The modified view of the importance of the century after 1750 is put forward in Mingay (1963b) and Chambers and Mingay (1966). A bibliography of Mingay's works is in his Festschrift, Holderness and Turner (1991). Sturgess (1966) argued for the clayland revolution, following on from Darby (1964), while Thompson (1968) argued for the 'second agricultural revolution'. Kerridge's arguments are set out in Kerridge (1967), in a more accessible form in Kerridge (1973), and as a specific attack on Mingay in Kerridge (1969b); the reply is Mingay (1969). Jones' view of the period 1650–1750 can be found in Jones (1965, 1967a, 1974 and 1981) and John's contributions in John (1960, 1961 and 1965). Allen's two agricultural revolutions are set out in Allen (1992) and summarised in Allen (1991); Clark's recent view is expressed in Clark (1993). Campbell and Overton (1993) argue that change accelerated after 1740. A revision of the Scottish experience is in Devine (1994).

Agricultural history is reviewed by Overton (1988). An earlier review is by Thirsk (1955), while Lennard (1964), Jones (1975), Bellamy, Snell and Williamson (1990) and Reed and Wells (1990a) are examples of future agendas for agrarian history past and present. An appreciation of the work of Hoskins is published in his Festschrift, Chalklin and Havinden (1974), together with a bibliography of his writings; Joan Thirsk's Festschrift is edited by Chartres and Hey (1990), which also contains an appreciation and bibliography. The Cambridge *Agrarian History* volumes covering the period of this book are those by Thirsk (1967d, 1984b and 1985b) and Mingay (1989). Reviews of this project include those by Jones (1968), Wrigley (1987a), Habakkuk (1987), Overton (1986b and 1990b) and Thompson (1990a). For Tawney see Terrill (1973), and for the Hammonds, Saville (1988). The debate inspired by Brenner's (1976) article is reprinted in Aston and Philpin (1985). Overviews of European agrarian history include Abel (1980) and Slicher van Bath (1963), though both are rather dated (Abel's book was first published in 1935).

Sources for agrarian history are not discussed explicitly in this book, but introductions may be found in Thirsk (1955), Stephens (1969 and 1981), Grigg (1967b) and Edwards (1991). Creasey (1981) is a more general bibliographical guide to agrarian history.

The quotations in the chapter are from or quoted in: p. 1, Marshall (1787), I, 262–3; p. 3, Curtler (1909), 162–3; p. 5, Kerridge (1967), 13; p. 6, Jones (1965), 1; Mingay (1989), 953; Allen (1991), 236; p. 7, Clark (1993), 246.

Chapter Two: Farming in the sixteenth century

The major works on farming in the sixteenth century are those by Kerridge (1967 and 1973) and Thirsk (1967d). A number of general texts on the economy and society of the sixteenth century include useful sections on agriculture: Clay (1984), Coward (1988), Hill (1969), Holderness (1976b), Hoskins (1976), Palliser (1983) and Sharpe (1987). The most vivid descriptions of farming techniques are to be found in contemporary books and farming diaries. Fortunately several of these have been reprinted and are fairly easily accessible. Diaries and accounts include those edited by Alcock (1981), Brassley, Lambert and Saunders (1988), Fussell (1936), Lodge (1927) and Woodward (1984). Collections of transcriptions of probate inventories, such as those edited by Havinden (1965), Moore (1977), Reed (1988), Steer (1950) and Trinder and Cox (1980), include some analysis of farming items. Definitions of unfamiliar terms may be found in Adams (1976), Kerridge (1967) and Worlidge (1704). Descriptions of farming operations are found in reprints of farming classics including Tusser (1984) for the sixteenth century, and, for the later period, Lisle (1757). Contemporary descriptions are also found in Thirsk and Cooper (1972), and in Lennard (1932). Farming techniques are described by Thirsk (1967b), Trow-Smith (1951) and Fussell (1965). Early editions of modern farming text books such as Fream (1892) and Watson and More (1924) are useful for understanding the basic principles and operations of farming, as are oral histories such as Evans (1960, 1966 and 1969). The gender division of labour is discussed in Clark (1919). Specific illustrations of farming activity for the middle ages can be found in Backhouse (1989) and, for the nineteenth century, in Jewell (1965), while Prince (1988) and Fussell (1984) discuss the depiction of agriculture in art. Vince (1982) is one of the most charming books on the material culture of farming. The problems of soil fertility in an historical context are discussed by Shiel (1991). The impact of the weather is considered by Bowden (1967a, pp. 45–62), Jones (1964a and 1975), Overton (1989b) and Stratton (1978). Varieties of farm economy are illustrated by Bowden (1967a) and the relationship between grain output and grain price is explored by Wrigley (1987c). Dyer (1995) discusses the extent of self-sufficiency in the middle ages, but there is little work on the early modern period.

Field systems are introduced in Yelling (1977), regional varieties are considered in Baker and Butlin (1973); and Kerridge (1992) deals specifically with commonfields. The single open-field village of Laxton is described in Orwin and Orwin (1938) and in Beckett (1989a). Examples of village bye-laws (although mostly from the middle ages) are given in Ault (1972); another example is included in the Reports of the Historical Manuscripts Commisson (1911). A useful typology of field systems is provided by Campbell (1981). There is a large literature on the origins of open-fields: a recent review with yet another new theory is by Fenoaltea (1988), but see also Dodgshon (1980), Dahlman (1980) and Rowley (1981). More references are given in the *Further reading* to Chapter Four.

The clearest exposition of the basis of tenures and estates is Riddall (1988, 7–49); but the topic is covered in other legal histories such as Baker (1979). Kerridge's (1969a) criticism of Tawney (1912) is important but rather difficult; see also Baker (1978) on the legal standing of customary tenants. Campbell (1942, 105–55) is helpful on the nature of freehold and copyhold. The geography of variations in landholding was quantified by Tawney (1912); regional variations are described in the Cambridge *Agrarian History*, Thirsk (1967d and 1984b), and in Bettey (1982), Drury (1987) and Hoyle (1987).

Laslett (1983), Sharpe (1987), Thomas (1971, 3–24) and Wrightson (1982) provide introductions to the social history of the sixteenth century; attitudes to work are discussed in Coleman (1956) and Thomas (1964); while religious and secular festivals are covered in Hutton (1994). Examples of dual employment are provided by Frost (1981), Hey (1969) and Spenceley (1973). The issue of status is discussed by Wrightson (1982 and 1986), and kin relationships by Cressy (1986). Estimates of proportions in the various status groups are given in Cornwall (1988). The gentry receive specific attention from Heal and Holmes (1994); the yeoman from Campbell (1942) and Marshall (1980); the parson from Brooks (1948); servants from Kussmaul (1981); and labourers from Everitt (1967a). The Norfolk 'labourers' in Smith (1989) are particularly interesting but probably untypical. Examples of relationships between social groups in agriculture can be found in Holderness (1976a) and Overton (1985). Tithes are discussed by Venn (1933, 150–82) and by Evans (1976). The literature on 'peasants' is vast. Wolf (1966) provides a good introduction to the subject; Beckett (1984b) and Neeson (1993) discuss the use of the term by historians. Macfarlane's thesis is set out in Macfarlane (1978 and 1987).

The first major study of an historical community was by Hoskins (1957), on Wigston Magna in Leicestershire, although his views on the Midland village community were set out earlier in Hoskins (1949). The genre of historical 'community studies' took off in the 1970s with Spufford (1974), on three Cambridgeshire villages; Hey (1974), for Myddle in Shropshire; and Wrightson and Levine (1979), on Terling in Essex. More recent examples include Nair (1988), on Highley in Shropshire, and Wrightson and Levine (1991), on Wickham in County Durham. 'Community' is a difficult concept; for an introduction see Macfarlane (1977). Administrative units are discussed in Sylvester (1969, 149–89), the manor by Kerridge (1969a), and examples of the relationship between manor and parish are provided by Campbell (1986).

There is a large geographical literature on regions: for a general overview see Grigg (1967a); for regions in an agricultural context see Chisholm (1964) and Tarrant (1974). Early ideas on agricultural regions are discussed in Darby (1954), Hoskins (1954) and Butlin (1990). The typology of *pays* comes from Everitt (1979), as modified and mapped by Thirsk (1987). A more recent description of English *pays* based on river basins is by Phythian-Adams (1993). Maps of population density are in Sheail (1972), and the distribution of building materials and architectural features can be found in Brunskill (1978). Contemporary descriptions of variations in the countryside are summarised in Taylor (1936). Thirsk's farming regions are set out in Thirsk (1967c and 1984b) and reviewed in Thirsk (1987). For an alternative regional structure see Kerridge (1967) and the comparison by Minchinton (1971–2). The cluster analysis is from Overton (1983a), but more sophisticated farm classifications have been undertaken for the middle ages by Power and Campbell (1992). The relationship between industry and agriculture is developed in Thirsk (1961) and summarised in Thirsk (1973b). Examples of the relationships between the two include Frost (1981), Hey (1969), Spenceley (1973) and Thirsk (1973a). The protoindustrialisation model is reviewed by Butlin (1986) and Clarkson (1985); the original model was set out by Mendels (1972) and extended by Medick (1976). The spatial pattern of nonconformity in the seventeenth century was revealed by Everitt (1970). Manning (1988) discusses the relationship between farming regions and social disorder, Ingram (1987) sexual behaviour and Underdown (1979 and 1985) politics, religion and sport. The issue of 'open' and 'close' parishes is reviewed by Banks (1988). Criticism of Thirsk's farming regions is in Overton (1983a and 1984c); Morrill (1987) criticises Underdown; the best critical survey of the whole issue is by Davie (1991).

There are many studies of particular rural economies: sheep-corn husbandry is described for East Anglia by Allison (1957) and Bailey (1990); and for Wiltshire by Kerridge (1953). Wood-pasture economies are described by Chalklin (1962), Evans (1984) and Zell (1985 and 1994); Pettit (1967) and Tubbs (1965) describe forest economies; and Ravensdale (1974) and Thirsk (1953) describe fenland farming. Studies which contrast a variety of agricultural regimes include Bettey (1977a) on the West Country, Glennie (1988a and 1988b) on Hertfordshire, Hoskins (1945) on Leicestershire, Chalklin (1965) on Kent, Thirsk (1957) on Lincolnshire, Overton (1991) on Norfolk and Suffolk, Cornwall (1954) on Sussex, Kerridge (1959) on Wiltshire, and Yelling (1969, 1970 and 1973) on Worcestershire. See also the references to changes in regional farming for Chapter Three.

The quotations in the chapter are from or quoted in: p. 19, Fussell (1936), 176; p. 32, Kent (1796), 28; p. 37, Macfarlane (1976), 553, Thompson (1991a), 372; p. 40, Ellis (1840), xii; p. 41, Thirsk and Cooper (1972), 170–1; p. 43, Macfarlane (1978), 163; p. 45, Hoskins (1949), 79, 72; p. 46, Hoskins (1957), 95; p. 47, Hoskins (1954), 5; and p. 50, Britton (1847), 11.

Chapter Three: Output and productivity, 1500–1850

Issues of changing agricultural production are of central concern to most agrarian histories written in the 'cows and ploughs' tradition. Basic economic concepts are explained in Ritson (1977), while Tivy (1990) is helpful on matters of agricultural ecology. Gregory King's forecasts can be found in Thirsk and Cooper (1972, 775), and more recent estimates in Wrigley and Schofield (1981). The sources for price information are to be found in the section on *Sources for tables* on p. 208. Agricultural prices are discussed by Bowden (1967a and 1985a), and Holderness (1989) in the Cambridge *Agrarian History*. More general discussions of prices are to be found in Outhwaite (1969) and O'Brien (1985).

Many of the arguments in this chapter are summarised in Overton (1996). The concepts of output and productivity are discussed in Overton and Campbell (1991), the importance of the difference between net and gross yields is pointed out by Slicher van Bath (1963), and by Wrigley (1987c). The specific sources for the estimates here are found in the list of sources for Tables 3.5–3.9. Output estimates based on population totals are found in Deane and Cole (1967); the demand equation method was pioneered by Crafts (1985) and developed by Jackson (1985). The difficulties of calculating the volume of output are discussed in Chartres (1985) and Holderness (1989). Land productivity estimates based on wage rates have been calculated by Clark (1991) who also provides benchmark figures for *c.* 1300 and 1850 from other sources (1991a). Indirect estimates of crop yields from probate inventories were first calculated by Overton (1979), adopted by Glennie (1988b), criticised by Allen (1988a), improved by Overton (1990a and 1991), and subjected to further development by Glennie (1991). Young's evidence on yields is treated favourably by Allen and Ó Gráda (1988), although not all historians would agree with their verdict. Turner (1982) presents evidence for crop yields *c.* 1800, which are commented on by Overton (1984a); Kain (1986) provides figures for the 1830s, and Craigie (1883) for the mid-nineteenth century. A long–term view of Norfolk yields from 1250 to 1854 can be found in Campbell and Overton (1993). Evidence of livestock productivity is given by Armitage (1980), Clutton-Brock (1982), Fussell (1929), Russell (1986), and for Scotland, by Gibson (1988). The Norfolk evidence is from Overton and Campbell (1992).

National estimates of labour productivity are provided by Clark (1991a), Overton (1990b) and Wrigley (1985). Total factor productivity in agriculture is discussed by Crafts (1987), Hueckel (1981) and McCloskey (1981), and reviewed by Mokyr (1987). Two very recent attempts by econometric historians to estimate output and productivity are those by Clark (1993) and Allen (1994). Pounds (1973) shows the possibilities for 'bottom-up' estimates of labour productivity.

The contribution of overseas trade in agricultural products is dealt with in Barnes (1930), Chartres (1985), John (1976), Ormrod (1985) and Thomas (1985a and 1985b). Landscape reclamation in general is covered in Darby (1951 and 1973a), Kerridge (1967) and Prince (1989). Holderness (1988) and Feinstein (1978 and 1988) provide information on captial expenditure. Marshlands and woodlands are meticulously plotted by Wilcox (1933). For the reclamation of fenlands and marsh see Darby (1983), Lindley (1982), Sheppard (1957) and Williams (1970a); for woodlands, see Darby (1951) and Rackham (1980 and 1986). The example of Wychwood comes from Belcher (1863); see also Emery (1974, pp. 158–62). Hoyle (1992a) and Thirsk (1992) examine the role of the crown in reclamation during the sixteenth and seventeenth centuries. Williams (1970b) deals with wasteland reclamation in England and Wales during the eighteenth and nineteenth centuries; he also provides a county case-study (1972), as does Hoskins (1943).

General changes in husbandry from 1650 to 1750 are covered in Thirsk (1984b). In addition see Brigg (1962 and 1964), Campbell and Overton (1993), Cornwall (1954 and 1960), Fieldhouse (1980), Hoskins (1951), Kenyon (1955), Large (1984), Long (1960), Longman (1977), Overton (1991), Pickles (1981), Skipp (1970) and Yelling (1973). Overton (1984c) discusses the use of probate inventories as a source for farming statistics; later sources are discussed in Grigg (1967b), Minchinton

(1953), Kain and Prince (1985), Dodd (1987) and Coppock (1984). The introduction of new fodder crops was first dealt with by Garnier (1896). Subsequent articles include Fussell (1955, 1959 and 1964), and Harvey (1949); more recent studies include one by Overton (1985) which traces the diffusion of turnips and clover in Norfolk and Suffolk, Emery (1976) which discusses the spread of clover in Wales, and Large (1984) for north Warwickshire. Examples of the improvements of light lands are provided by Jones (1960), Harris (1961), Grigg (1966), and Campbell and Overton (1993). Thirsk (1983 and 1985a) writes about the spread of agricultural information more generally, and (1974) on the diffusion of tobacco cultivation in England. Kerridge (1967) also deals with crop innovations from the sixteenth to the eighteenth century, while later introductions are dealt with by Morgan (1989). Exemplars across the Channel are introduced (in English) by Bieleman (1991). The replacement of rye was first discussed by Ashley (1928) and the history of the potato is covered extensively in a classic book by Salaman (1949).

The concepts behind regional specialisation are covered in Kussmaul (1990) which also provides the empirical evidence of specialisation as evidenced from marriage seasonality. In addition to the material in the *Cambridge Agrarian History*, inventory studies indicating specialisation include Broad (1980), Campbell and Overton (1993), Edwards (1978), Glennie (1988a), Overton (1985 and 1991), Overton and Campbell (1992) and Yelling (1970 and 1973).

The literature on pests and diseases has been discussed in the *Further reading* for Chapter Two. Determinants of yields are discussed by Overton (1991). Little has been written on early plant breeding, but see Allen (1992), Overton (1991) and Pusey (1839). For the importance of nitrogen see Chorley (1981), Russell (1913, and many more editions), Shiel (1991) and Overton (1991) which deals with the impact of the Norfolk four-course rotation. The evidence of increased cultivations comes from Glennie (1988b). The importance of lime is considered by Havinden (1974); marl by Mathew (1993) and, using examples from Norfolk, by Prince (1964). Underdraining in the nineteenth century is covered by Phillips (1989). Woodward (1990) describes the varieties of manures used by farmers in the early modern period. A 'second agricultural revolution' was claimed by Thompson (1968) citing evidence of the import of feedstuffs and fertilisers. Grassland improvements are discussed by Lane (1980); there is also some useful material in Carrier (1936) and Bedford Franklin (1953). Contemporary writers on clover include Hartlib (1651), Blith (1652), Yarranton (1663) and Worlidge (1697). Watermeadows are discussed in Bettey (1977b), Bowie (1987b), Kerridge (1953), and Wade-Martins and Williamson (1994).

The classic work on livestock is Trow-Smith (1957 and 1959); a more lavish introduction is given in Hall and Clutton-Brock (1989). Bakewell's biography is by Pawson (1957), although more up-to-date material can be found in Russell (1986). For cattle, Russell (1981) discusses the longhorn, and Walton (1984) charts the diffusion of the shorthorn and, in an important article (1986), assesses the contribution of improved cattle to productivity. The most comprehensive book on sheep is by Ryder (1983), preceded by an earlier article (1964); the development of new breeds is discussed in Bowie (1987a) and Copus (1989); the spread of new breeds using farm sale advertisements is charted by Walton (1983) for Oxfordshire, and by Wade-Martins (1993) for Norfolk and Suffolk. Pigs have their own history in Wiseman (1986). Ley husbandry or 'up and down' husbandry is described by Kerridge (1967) and reviewed by Broad (1980) and Allen (1992). The Norfolk four-course rotation is described by Shiel (1991) and the roles of turnips and clover are also considered by Overton (1985 and 1991).

Collins (1981) provides an excellent review of the introduction of labour-saving machinery into English agriculture and Brown (1989) provides plenty of illustrations. The classic work on the plough is Passmore (1930), and Marshall (1978) looks at the development of the Rotheram plough. Tull is discussed by Fussell (1973) and Hidden (1989), while the diffusion of drill husbandry is considered by Wilkes (1981). The switch from the sickle to the scythe is discussed in Collins (1969), but his conclusions are questioned by Perkins (1976 and 1977); there is a further contribution by Roberts (1979), while Morgan (1975) discusses nineteenth-century harvesting more generally. The spread of the threshing machine is covered in Collins (1972) and Macdonald (1975). Details of the innovation pattern for a number of machines are given in Walton (1973 and 1979).

Hunt (1967) discusses the significance of nutrition for labour productivity. Langdon (1986) writes about horses and oxen in the middle ages but includes much material of relevance to later periods. Wrigley (1988 and 1991) contributes on energy availability and labour productivity before 1800, and (1986) discusses employment on the land more generally in the nineteenth century (including a review of census evidence for 1831 and 1851). The development of steam power in agriculture is reviewed by Spence (1960) and further references on steam power, and on tools and machinery in general, are given in the exhaustive bibliography by Morgan (1984). The relationship of labour productivity to farm size is discussed by Allen (1988b, 1991 and 1992), but see also Clark (1991a and 1991b). The labour requirements of the Norfolk four-course rotation are illustrated by Timmer (1969). Changes in employment are outlined by Armstrong (1989) and Kussmaul (1981). Pioneering use of labour accounts, though not explicitly for the calculation of labour productivity, is by Smith (1989) and Pounds (1973).

Perkins (1939) is the best bibliography of books on English farming before 1900. The work of the Georgical Committee is considered by Lennard (1932), and by Thirsk and Cooper (1972). The Board of Agriculture is described by Mitchison (1959); the *General Views* are listed in McGregor (1961) and Perkins (1939). There is much material on Arthur Young, including extracts from his works in Mingay (1975) and a biography by Gazley (1973). Young's rival, Marshall, is praised in Kerridge (1968), and Horn (1982) provides a short biography. Young's empirical ethos is illustrated in Young (1771), and Marshall's in Marshall (1778 and 1779). Pretty (1991) includes several examples of contemporary agricultural experiments. The development of agricultural societies and agricultural education is dealt with by Fox (1979), Sykes (1981), Hudson (1972), Goddard (1988, 1989 and 1991) and Wilmot (1990). Estate administration in the early modern period is covered in Hainsworth (1992) and Beckett (1990a). Pollard (1965) mentions agricultural accounting but there is little other literature on the topic; see Colyer (1975).

The quotations in the chapter are from or quoted in: p. 90, quoted in Lindley (1982), 7; p. 91, Scarfe (1988), 175; pp. 91–2, Anon. (1752), 502; p. 92, quoted in Williams (1970a), 57; p. 102, quoted in Salaman (1949), 507; pp. 104–5, Lucas

(1892), 205; p. 106, quoted in Pretty (1991), 145; p. 107, Hartlib (1651), 15; p. 110, Tusser (1984), 44; Blith (1652), 184; p. 111, Russell (1913), 58; p. 112, quoted in Palliser (1976), 103; p. 117, Moore (1946), 17; p. 129, Grigg (1967b), 76; Public Record Office (1975), 3; Lucas (1892), 192–3; and p. 130, Pusey (1850), 438.

Chapter Four: Institutional change, 1500–1850

Marketing in England from 1500 to 1850 is a relatively under-researched field, and most recent work has been for the middle ages and for Europe, for example Campbell *et al.* (1993), Britnell (1993), Dyer (1992) and Grantham (1989a and 1989b). A good general study of primitive markets is by Hodges (1988), while Kerridge (1986 and 1988) provides introductions to early modern markets. The four major works on agricultural marketing covering 1500–1850 are Gras (1915), Everitt (1967b), Chartres (1985) and Perren (1989). Outhwaite (1981 and 1991) considers public policy, middlemen are reviewed by Westerfield (1915), and Hey (1980) provides an example of the local history of marketing. A general overview at a theoretical level inspired by the works of Polanyi can be found in Dodgshon (1987 and summarised in 1990). The market regulations cited in the chapter are listed in Historical Manuscripts Commission (1907, 129–30). The provisioning of London is dealt with by Fisher (1935) and Chartres (1986); Baker (1970) and Thwaites (1985 and 1991) give local examples of marketing in Kent and Oxfordshire. Bowden (1962) includes an account of the wool market, Edwards (1988) of the horse trade; Edwards (1981), Fussell and Goodman (1936) and Blackman (1975) describe the cattle trade and the traffic in livestock to London; and Woodward (1973) the Irish cattle trade. Droving is specifically considered by Bonser (1970) and Woodward (1977). The relationship of a national economy to subsistence crises is discussed by Walter and Schofield (1989a). Subsistence crises are also considered by Wrigley and Schofield (1981, Appendix 10) and famine by Appleby (1973 and 1978). The transport system is explicitly considered by Chartres (1977), and Frearson (1994) for the sixteenth and seventeenth centuries. Examples of individual farmers' market activities can be found in Alcock (1981), and Brassley, Lambert and Saunders (1988), and an example of distribution of buyers and sellers at White Down fair is in Hamer (1968). Many contributions to the debate on the 'moral economy' (see below) discuss marketing, particularly Stevenson (1985). The overseas trade in corn is covered in Barnes (1930), Ormrod (1985) and Chambers and Mingay (1966); and the impact of the corn laws on domestic production is covered by John (1976) and Fairlie (1969).

In contrast to the literature on markets that on enclosure is vast. The classic works are by Slater (1907), Gonner (1912) and Curtler (1920), which are all still worth reading; Yelling (1977) is a more modern account which considers pre-parliamentary enclosure. The figures on the chronology of enclosure come from Wordie (1983), although his estimates have been criticised. The chronology of enclosure between 1600 and 1750 is also reviewed by Butlin (1979). Thirsk (1967a) gives an overview of sixteenth-century enclosure; Beresford (1961) describes the process of pre-parliamentary enclosure; more specific studies are by Reed (1981 and 1984a) on the enclosure chronology of Buckinghamshire before parliamentary enclosure. Local evidence is also provided in Allen (1992), Beresford (1948) and

Hodgshon (1979). Details of the chronology and geography of parliamentary enclosure are in Tate (1978), and summarised in Turner (1980); there are numerous published examples of individual enclosures: see the references in Blum (1981), Turner (1984) and Brewer (1972).

General changes in tenures are discussed in Bowden (1967a), Clay (1981 and 1985) and Beckett (1989b). The classic argument over the development of lease-hold is between Tawney (1912) and Kerridge (1969a); Allen (1992) is an important recent contribution, but the best discussions of the issue are by Gregson (1989) and Hoyle (1990). Appleby (1975), Hoyle (1987), Spufford (1974), and Wrightson and Levine (1979) also have useful material. The example of Wigston Magna comes from Hoskins (1957), Cotesbach from Parker (1948), and the activities of the Verneys are chronicled by Broad (1990). Another example is given in Thorpe (1965) on Wormleighton in Warwickshire. Fenland enclosures are dealt with in Lindley (1982). Turner's book (1980), pamphlet (1984) and short article (1989) are the main introductions to parliamentary enclosure, together with the survey article by Blum (1981), though Allen's recent book (1992) makes important new contri-butions. Tate (1967) describes how parliamentary enclosure was actually carried out.

The impact of enclosure on the landscape is outlined by Hoskins (1955) and Taylor (1975); examples of particular counties can be found in the volumes of the *Making of the English Landscape* series, including Bigmore (1979) on Bedfordshire and Huntingdonshire, Steane (1974) on Northamptonshire, Emery (1974) on Oxfordshire, and Palliser (1976) on Staffordshire. Explanations of the timing and rate of enclosure can be found in Bowden (1952 and 1967a) and Martin (1988) for the sixteenth century; and for parliamentary enclosure in Chambers and Mingay (1966), Crafts (1977), McCloskey (1972, 1975a 1975b and 1989), Purdum (1978) and Turner (1981b). The relationship between enclosure and output is discussed in Allen (1992), Havinden (1961) and Turner (1986); local examples are provided in Grigg (1966) and Harris (1961).

For landownership in general see Clay (1985), Mingay (1963a) and Thompson (1963). Estimates of landownership are discussed by Cooper (1967) and Thompson (1966 and 1969); the figures used here are those adopted by Mingay (1976). The dissolution of the monasteries is described in Youings (1971), while Swales (1966) provides a Norfolk example. The rise of the gentry is reviewed by Mingay (1976), the original material includes Tawney (1941), and a rebuttal by Trevor-Roper (1953); the debate was further complicated by Stone (1965). Local studies demon-strating the rise of the gentry include Everitt (1966) and Blackwood (1978). Bateman (1883) is the classic work on landownership in the late nineteenth century.

For the effects of the civil wars see Habakkuk (1965). The thesis of the rise of the great estates is set out by Habakkuk (1940, restated 1979–81, and somewhat modified 1994). Criticisms and further contributions are to be found in Allen (1992), Beckett (1977 and 1984a), English and Saville (1983), Holderness (1974), and Stone and Stone (1984). Two of the most famous agricultural estates in England are chronicled by Parker (1975) and Wade-Martins (1980), for Holkham in Norfolk, and Rosenheim (1989), for Raynham in Norfolk. The varying fortunes of farmers in the seventeenth century are reviewed by Thirsk (1970a).

Social differentiation is reviewed in Wrightson (1977). The Cambridgeshire evidence is from Spufford (1974); there are also important contributions from Allen (1992), Mingay (1962) and Wordie (1974 and 1982). The runs of bad harvests are identified in Hoskins (1964 and 1968). Lavrovsky (1956) illustrates the variety of experience of social differentiation in three villages. For discussion of the nineteenth-century data on farm sizes see Grigg (1967b and 1963); distribution maps are in Overton (1986a).

Marx ascribed the creation of the proletariat to the decline of the peasant at parliamentary enclosure and his argument was broadly followed by Hammond and Hammond (1912). Collins (1967) provides a detailed textual rebuttal of Marx, albeit using an inappropriate Popperian methodology: equally inappropriate is Lazonick's (1974) verbatim acceptance of Marx. The revisionist view comes from Chambers (1952), and later Mingay (1968). In turn this view is criticised by Snell (1985), and Neeson (1984, 1989 and 1993), who argues for the decline of the commoner with parliamentary enclosure. Humphries (1990) and King (1989) also contribute to the discussion. The survival of family farms into the nineteenth century is stressed by Reed (1990). The figures of the number of labourers are from Tawney (1912) and Cornwall (1988). The ratio between labourers and occupiers not employing labour in 1831 was discussed by Saville (1969). The decline of servants is covered in Kussmaul (1981) and Short (1984).

Social relations in general are reviewed in Malcolmson (1981). Landlord–tenant relations are discussed in Beckett (1977 and 1987); the 'agricultural depression' is discussed by Mingay (1956) and Beckett (1982); and the lease covenant is evaluated by Macdonald (1976). Tenant right is treated by McQuiston (1973) and Fisher (1983). The game laws are dealt with in Munsche (1981), and also in Thompson (1975). Thompson (1974 and 1991a) also argues for the decline of paternalism. Armstrong (1981, 1988, 1989 and 1990) rather dominates writing on nineteenth-century farm labourers, but see also Jones (1964b). Living standards are discussed in the Cambridge *Agrarian History*, by Everitt (1967a), and Armstrong (1989); regional changes are mapped in Hunt (1986). Local evidence is provided in a series of studies by Richardson (1976, 1991 and 1993). The literature on the poor law is vast, but introductions are to be found in Armstrong (1989), Marshall (1985), Rose (1972), and Slack (1990). The settlement laws are considered by Landau (1988) and Snell (1985). The changing nature of women's work is discussed in Clark (1919), Hill (1989), Roberts (1979), Snell (1981 and 1985, Chapter 1) and Valenze (1991). The impact of the English poor law on economic development has been reviewed recently by Solar (1995).

The most straightforward guide to popular unrest throughout the period is Charlesworth (1983). Manning (1988) and Walter and Wrightson (1976) are more general studies of the early period; for the later period see Stevenson (1979) and Wells (1988 and 1990). Particular studies include MacCulloch (1979) on Kett's rebellion of 1549, Walter (1985) on the Oxfordshire rising of 1596, Martin (1983) on the Midland Revolt of 1607, Sharp (1980) on the western rising of 1626–32, Lindley (1982) on fenland riots, Wells (1988) and Stevenson (1974) on riots in the late eighteenth century, Bohstedt (1983) on Devon in 1795 and 1800–1, Neeson (1984) on opposition to parliamentary enclosure in Northamptonshire, Peacock (1963) on East Anglia in the nineteenth century, and Hobsbawm and Rudé (1969) and Charlesworth (1979) on Captain Swing. The notion of the 'moral economy' as set out in Thompson (1971) has come in for considerable criticism, for example by Williams (1984), and modification, for example by Bohstedt (1992). Thompson is defended by Charlesworth and Randall (1987) amongst others, but has himself published a comprehensive reply to his critics, (1991b). A recent review is by Wells (1994).

The quotations in the chapter are from or quoted in: p. 139, Clare (1964), 76–7; p. 144, quoted in Baker (1970), 138; p. 159, quoted in Yelling (1977), 120; quoted in Steane (1974), 233; p. 167, Kent (1796), 73; Lucas (1892), 282; p. 170, quoted in Hill (1961), 201; p. 177, Palliser (1976), 128; Young (1808), 32–3; Hoskins (1957), 249; p. 185, Jenkins (1869), 473; and pp. 190–1, Thompson (1991b), 188.

Chapter Five: The agricultural revolution reconsidered

Mid-nineteenth-century agriculture is described in Holderness (1981), Jones (1962), Orwin and Whetham (1964) and Thompson (1968). An overview of regions in England from 1600 to 1914 is in Butlin (1990). County by county descriptions of nineteenth-century farming are found in the 'prize essays' in the *Journal of the Royal Agricultural Society of England*; they are listed in McGregor (1961, cii–ciii). High farming is discussed specifically by Holderness (1991), and by Scott Watson and Hobbs (1937, 87–113). Contemporaries' depictions of regional differences are reviewed by Prince (1989), while Overton (1986a) provides maps of farming practice for Great Britain during the nineteenth century, and Atkins (1988) considers the impact of London on production patterns. Goldstone (1988) relates the natural environment to agricultural innovation and regional differentiation, but is criticised by Hopcroft (1994). The example of the differences between Yorkshire and Wessex is from Bowie (1990).

The arguments for a Malthusian check *c.* 1650 are put by Schofield (1983), and also by Palliser (1982). Boserup (1965 and 1981) gives the key statements of her views on the positive effects of population pressure, but there is an extensive literature on the relationships between population growth and agricultural change: Grigg (1979, 1980 and 1982) provides good introductions, and Lipton (1990) a sophisticated overview. For an illuminating small-scale study of the impact of population pressure see Skipp (1978).

Jones' view of the period 1650–1750 can be found in Jones (1965, 1967a and 1974) and John's contributions in John (1960 and 1965). Criticism of their views is in Flinn (1966), O'Brien (1977), Overton (1983b) and Glennie (1988b). The examples of price responsiveness are from Hueckel (1976), Macdonald (1980) and Copus (1989). General discussions of farmers' responses to prices can be found in Giles (1956). Innovation as a form of risk aversion is discussed in Overton (1989b).

Perhaps the best starting point on capitalism in England is Holton (1985), followed by Tribe (1981, Chapter Two) who deals specifically with agrarian capitalism, as does Saville (1969). Goodman and Redclift (1981) provide a broader context. The 1950s debate on the transition to capitalism is reprinted in Hilton (1976); more recently, Brenner's arguments were set out in Brenner (1976) and again as a reply to criticism in (1982). Both articles are reprinted in Aston and Philpin (1985), along with the comments on Brenner's thesis published in *Past and*

Present. Other comments and criticisms on Brenner can be found in Medick (1981), Searle (1986), Hagen (1988), Glennie (1988c) and Hoyle (1990). The geography of agrarian capitalism is considered in a book (1987) and an essay (1990) by Dodgshon. The argument for a changing attitude to farming based on contemporary literature is set out in Tribe (1978).

The quotations in the chapter are from or quoted in: p. 194, cited in Holderness (1991), 154; p. 196, Caird (1852), 476; p. 203, Overton (1989b), 85; and p. 206, quoted in Tribe (1981), 77.

Bibliography

Abbreviations

AH   Agricultural History
AHR   Agricultural History Review
EcHR   Economic History Review, second series
EEcH   Explorations in Economic History
JEcH   Journal of Economic History
JHG   Journal of Historical Geography
JRASE   Journal of the Royal Agricultural Society of England
P&P   Past and Present
TIBG   Transactions of the Institute of British Geographers

The place of publication is London unless otherwise stated.

Abel, W. 1980. *Agricultural fluctuations in Europe from the thirteenth to the twentieth centuries*, trans O. Ordish.

Adams, I.H. 1976. *Agrarian landscape terms: a glossary for historical geography*, Institute of British Geographers, Special Publication 9.

Alcock, N.W. (ed.) 1981. *Warwickshire grazier and London skinner 1532–1555: the account book of Peter Temple and Thomas Heritage,* Records of Social and Economic History, new series 4.

Allen, R.C. 1988a. Inferring yields from probate inventories, *JEcH*, 48: 117–25.

Allen, R.C. 1988b. The growth of labour productivity in early modern English agriculture, *EEcH*, 25: 117–46.

Allen, R.C. 1991. The two English agricultural revolutions, 1459–1850. In Campbell and Overton (1991), 236–54.

Allen, R.C. 1992. *Enclosure and the yeoman: the agricultural development of the South Midlands 1450–1850*, Oxford.

Allen, R.C. 1994. Agriculture during the industrial revolution. In Floud and McCloskey (1994), I, 96–122.

Allen, R.C. and Ó Gráda, C. 1988. On the road again with Arthur Young: English, Irish, and French agriculture during the industrial revolution, *JEcH*, 48: 93–116.

Allison, K.J. 1957. The sheep-corn husbandry of Norfolk in the sixteenth and seventeenth centuries, *AHR*, 5: 12–30.

Anderson, B.L. and Latham, A.J.H. (eds.) 1986. *The market in history.*

Anon. 1752. Management of three farms in the county of Norfolk, *Gentleman's Magazine*, 22: 502.

Appleby, A.B. 1973. Disease or famine? Mortality in Cumberland and Westmorland, 1580–1640, *EcHR*, 26: 403–32.

Appleby, A.B. 1975. Agrarian capitalism or seigneurial reaction? The northwest of England, 1500–1700, *American Historical Review*, 80: 574–94.

Appleby, A.B. 1978. *Famine in Tudor and Stuart England*, Liverpool.

Armitage, P.L. 1980. A preliminary description of British cattle from the late twelfth to the early sixteenth century, *Ark*, 7: 405–13.

Armstrong, W.A. 1981. The workfolk. In Mingay (1981), II, 491–505.

Armstrong, W.A. 1988. *Farmworkers: a social and economic history.*

Armstrong, W.A. 1989. Labour I: rural population growth, systems of employment, and incomes. In Mingay (1989), 671–95.

Armstrong, W.A. 1990. The countryside. In Thompson (1990b), 87–154.

Ashley, W. 1928. *The bread of our forefathers: an enquiry into economic history*, Oxford.

Aston, T.H., Coss, P.R., Dyer, C. and Thirsk, J. (eds.) 1983. *Social relations and ideas: essays in honour of R.H. Hilton.*

Aston, T.H. and Philpin, C.H.E. (eds.) 1985. *The Brenner debate: agrarian class structure and economic development in pre-industrial Europe*, Cambridge.

Atkins, P.J. 1988. The charmed circle: Von Thünen and agriculture around nineteenth-century London, *Geography*, 72: 129–39.

Ault, W.O. 1972. *Open-field farming in medieval England.*

Avery, B.W., Findlay, D.C. and Mackney, D. 1974. *Soil map of England and Wales*, Soil Survey of England and Wales.

Backhouse, J. 1989. *The Luttrell Psalter.*

Bailey, M. 1990. Sand into gold: the evolution of the foldcourse system in west Suffolk, 1200–1600, *AHR*, 38: 40–57.

Baker, A.R.H. and Butlin, R.A. (eds.) 1973. *Studies of field systems in the British Isles*, Cambridge.

Baker, A.R.H. and Gregory, D. (eds.) 1984. *Explorations in historical geography: interpretive essays*, Cambridge.

Baker, A.R.H., Hamshere, J.D. and Langton, J.L. (eds.) 1970. *Geographical interpretations of historical sources*, Newton Abbot.

Baker, A.R.H. and Harley, J.B. (eds.) 1973. *Man made the land*, Newton Abbot.

Baker, D. 1970. The marketing of corn in the first half of the eighteenth century: north east Kent, *AHR*, 18: 126–50.

Baker, J.H. 1978. Agrarian changes and security of tenure, *Seldon Society*, 94: 180–7.

Baker, J.H. 1979. *An introduction to English legal history*, 2nd edn.

Banks, S.J. 1988. Nineteenth-century scandal or twentieth-century model? A new look at open and close parishes, *EcHR*, 41: 51–73.

Barnes, D.G. 1930. *A history of the English corn laws from 1660–1846*, New Haven. Reprinted New York 1961 and 1965.

Bateman, J. 1883. *Great landowners of England and Wales*, 4th edn.

Beckett, J.V. 1977. English landownership in the later seventeenth and eighteenth centuries: the debate and the problems, *EcHR*, 30: 567–81.

Beckett, J.V. 1982. Regional variation and the agricultural depression, 1730–50, *EcHR*, 35: 35–51.

Beckett, J.V. 1984a. The pattern of landownership in England and Wales, 1660–1880, *EcHR*, 37: 1–22.

Beckett, J.V. 1984b. The peasant in England: a case of terminological confusion?, *AHR*, 32: 113–23.

Beckett, J.V. 1987. *The aristocracy in England 1660–1914*, Oxford.

Beckett, J.V. 1989a. *A history of Laxton: England's last open field village*, Oxford.

Beckett, J.V. 1989b. Landownership and estate management. In Mingay (1989), 545–640.

Beckett, J.V. 1990a. Estate management in eighteenth-century England: the Lowther-Spedding relationship in Cumberland. In Chartres and Hey (1990), 55–72.

Beckett, J.V. 1990b. *The agricultural revolution*, Oxford.

Bedford Franklin, T. 1953. *British grasslands from the earliest times to the present day*.

Beier, A.L. and Finlay, R. (eds.) 1986. *London 1500–1700: the making of the metropolis*.

Belcher, C. 1863. On the reclaiming of waste lands as instanced in Wichwood Forest, *JRASE*, 24: 271–85.

Bellamy, L., Snell, K.D.M. and Williamson, T. 1990. Rural history: the prospect before us, *Rural History*, 1: 1–4.

Beresford, M.W. 1948. Glebe terriers and open-field Leicestershire. In Hoskins (1948), 77–126.

Beresford, M.W. 1961. Habitation versus improvement: the debate on enclosure and agreement. In Fisher (1961), 40–69.

Bettey, J.H. 1977a. *Rural life in Wessex 1500–1900*. Republished Gloucester 1987.

Bettey, J.H. 1977b. The development of water meadows in Dorset during the seventeenth century, *AHR*, 25: 37–43.

Bettey, J.H. 1982. Land tenure and manorial custom in Dorset, 1570–1670, *Southern History*, 4: 33–54.

Beveridge, W. 1939. *Prices and wages in England from the twelfth to the nineteenth century*.

Bieleman, J. 1991. Dutch agriculture in the golden age, 1570–1660, *Economic and Social History in the Netherlands*, Amsterdam, 4: 159–85.

Bigmore, P. 1979. *The Bedfordshire and Huntingdonshire landscape*.

Blackman, J. 1975. The cattle trade and agrarian change on the eve of the railway age, *AHR*, 23: 48–62.

Blackwood, B.G. 1978. *The Lancashire gentry and the Great Rebellion*, Manchester.

Blith, W. 1652. *The improver improved or the survey of husbandry surveyed*.

Blum, J. 1981. English parliamentary enclosure, *Journal of Modern History*, 53: 477–504.

Bohstedt, J. 1983. *Riots and community politics in England and Wales, 1790–1810*, Cambridge, Mass.

Bohstedt, J. 1992. The moral economy and the discipline of historical context, *Journal of Social History*, 26: 265–84.

Bonfield, L., Smith, R. and Wrightson, K. (eds.) 1986. *The world we have gained: histories of population and social structure*, Oxford.

Bonser, K.J. 1970. *The drovers*.

Boserup, E. 1965. *The conditions of agricultural growth: the economics of agrarian change under population pressure*.

Boserup, E. 1981. *Population and technology*, Oxford.

Bowden, P.J. 1952. Movements in wool prices, 1490–1610, *Yorkshire Bulletin of Economic and Social Research*, 4: 109–24.

Bowden, P.J. 1962. *The wool trade in Tudor and Stuart England*.

Bowden, P.J. 1967a. Agricultural prices, farm profits, and rents. In Thirsk (1967d), 593–685.

Bowden, P.J. 1967b. Statistical appendix. In Thirsk (1967d), 814–70.

Bowden, P.J. 1985a. Agricultural prices, wages, farm profits and rents. In Thirsk (1985b), 1–118.

Bowden, P.J. 1985b. Statistical appendix. In Thirsk (1985b), 827–902.

Bowie, G. 1987a. New sheep for old – change in sheep farming in Hampshire, 1792–1879, *AHR*, 35: 15–24.

Bowie, G. 1987b. Watermeadows in Wessex: a re-evaluation for the period 1640–1850, *AHR*, 35: 151–8.

Bowie, G. 1990. Northern wolds and Wessex downlands: contrasts in sheep husbandry and farming practice, 1770–1850, *AHR*, 38: 117–26.

Brassley, P., Lambert, A. and Saunders, P. (eds.) 1988. *Accounts of the Reverend John Crakanthorp of Fowlmere 1682–1710*, Cambridgeshire Records Society 8, Cambridge.

Brenner, R. 1976. Agrarian class structure and economic development in pre-industrial Europe, *P&P*, 70: 30–74. Reprinted in Aston and Philpin (1985), 10–63.

Brenner, R. 1982. The agrarian roots of European capitalism, *P&P*, 97: 16–113. Reprinted in Aston and Philpin (1985), 213–327.

Brewer, J.G. 1972. *Enclosures and the open-fields: a bibliography*, Reading.

Brigg, M. 1962. The Forest of Pendle in the seventeenth century, part one, *Transactions of the Historic Society of Lancashire and Cheshire*, 113: 65–96.

Brigg, M. 1964. The Forest of Pendle in the seventeenth century, part two, *Transactions of the Historic Society of Lancashire and Cheshire*, 115: 65–90.

Britnell, R.H. 1993. *The commercialization of English society 1000–1500*, Cambridge.

Britton, J. (ed.) 1847. *The natural history of Wiltshire; by John Aubrey*.

Broad, J. 1980. Alternate husbandry and permanent pasture in the Midlands 1650–1800, *AHR*, 28: 77–89.

Broad, J. 1990. The Verneys as enclosing landlords, 1600–1800. In Chartres and Hey (1990), 27–54.

Brooks, F.W. 1948. The social position of the parson in the sixteenth century, *Journal of the British Archaeological Association*, 3rd series, 10: 23–37.

Brown, J. 1989. *Farm machinery 1750–1945*.

Brunskill, R.W. 1978. *Illustrated handbook of vernacular architecture*.

Butlin, R.A. 1979. The enclosure of open fields and extinction of common rights in England *circa* 1600–1700: a review. In Fox and Butlin (1979), 65–82.

Butlin, R.A. 1986. Early industrialization in Europe: concepts and problems,

Geographical Journal, 152: 1–8.

Butlin, R.A. 1990. Regions in England and Wales, *c.* 1600–1914. In Dodgshon and Butlin (1990), 223–54.

Caird, J. 1852. *English agriculture in 1850–51.*

Campbell, B.M.S. 1981. Commonfield origins – the regional dimension. In Rowley (1981), 112–29.

Campbell, B.M.S. 1986. The complexity of manorial structure in medieval Norfolk: a case study, *Norfolk Archaeology*, 39: 225–61.

Campbell, B.M.S., Galloway, J.A., Keene, D. and Murphy, M. 1993. *A medieval capital and its grain supply: agrarian production and distribution in the London region c. 1300*, Historical Geography Research Series 30.

Campbell, B.M.S. and Overton, M. 1993. A new perspective on medieval and early modern agriculture: six centuries of Norfolk farming, *c.*1250–*c.*1850, *P&P*, 141: 38–105.

Campbell, B.M.S. and Overton, M. (eds.) 1991. *Land, labour and livestock: historical studies in European agricultural productivity*, Manchester.

Campbell, M. 1942. *The English yeoman*, New Haven.

Cannon, J. *et al.* (eds.) 1988. *A dictionary of historians*, Oxford.

Carrier, E.H. 1936. *The pastoral heritage of Britain: a geographical study*, London.

Carus-Wilson, E.M. (ed.) 1954–62. *Essays in economic history*, 3 vols.

Chalklin, C.W. 1962. The rural economy of a Kentish Wealden parish, 1650–1750, *AHR*, 10: 29–45.

Chalklin, C.W. 1965. *Seventeenth-century Kent.*

Chalklin, C.W. and Havinden, M.A. (eds.) 1974. *Rural change and urban growth.*

Chaloner, W.H. and Ratcliffe, B.M. 1977. *Trade and transport: essays in economic history in honour of T.S. Willan*, Manchester.

Chambers, J.D. 1952. Enclosure and labour supply in the industrial revolution, *EcHR*, 5: 319–43. Reprinted in Glass and Eversley (1965), 308–27.

Chambers, J.D. and Mingay, G.E. 1966. *The agricultural revolution 1750–1880.*

Charlesworth, A. 1979. *Social protest in a rural society*, Historical Geography Research Series 1.

Charlesworth, A. (ed.) 1983. *An atlas of rural protest in Britain 1548–1900.*

Charlesworth, A. and Randall, A. 1987. Morals, markets and the English crowd in 1766, *P&P*, 114: 200–13.

Chartres, J.C. 1977. *Internal trade in England 1500–1700*, Studies in Economic and Social History, Basingstoke.

Chartres, J.C. 1985. The marketing of agricultural produce. In Thirsk (1985b), 406–502.

Chartres, J.C. 1986. Food consumption and internal trade. In Beier and Finlay (1986), 168–96.

Chartres, J.C. and Hey, D. 1990. *English rural society, 1500–1800: essays in honour of Joan Thirsk*, Cambridge.

Chisholm, M. 1964. Problems in the classification and use of farming-type regions, *TIBG*, 35: 91–103.

Chorley, G.P.H. 1981. The agricultural revolution in northern Europe, 1750–1880: nitrogen, legumes and crop productivity, *EcHR*, 34: 71–93.

Chorley, R.J. and Haggett, P. (eds.) 1967. *Models in geography.*

Clare, J. 1964. *The shepherd's calendar*, ed. Eric Robinson and Geoffrey Somerfield.

Clark, A. 1919. *Working life of women in the seventeenth century*, 3rd edn with an introduction by A.L. Erickson, 1992.

Clark, G. 1991a. Labour productivity in English agriculture, 1300–1860. In Campbell and Overton (1991), 211–35.

Clark, G. 1991b. Yields per acre in English agriculture 1250–1860: evidence from labour inputs, *EcHR*, 44: 445–60.

Clark, G. 1993. Agriculture and the industrial revolution, 1700–1850. In Mokyr (1993), 227–66.

Clark, P. (ed.) 1984. *The transformation of English provincial towns, 1600–1800*.

Clarkson, L.A. 1985. *Proto-industrialization: the first phase of industrialization?*

Clay, C. 1981. Life leasehold in the western counties of England 1650–1850, *AHR*, 29: 83–96.

Clay, C. 1984. *Economic expansion and social change: England 1500–1700*, 2 vols., Cambridge.

Clay, C. 1985. Landlords and estate management in England. In Thirsk (1985b), 119–251.

Clutton-Brock, J. 1982. British cattle in the eighteenth century, *Ark*, 9: 55–9.

Coleman, D.C. 1956. Labour in the English economy of the seventeenth century, *EcHR*, 8: 280–95. Reprinted in Carus-Wilson (1954–62), II, 291–308.

Coleman, D.C. and John, A.H. (eds.) 1976. *Trade, government and economy in pre-industrial England*.

Collins, E.J.T. 1969. Harvest technology and labour supply in Britain, 1790–1870, *EcHR*, 22: 453–73.

Collins, E.J.T. 1972. The diffusion of the threshing machine in Britain, 1790–1880, *Tools and Tillage*, 2: 16–33.

Collins, E.J.T. 1981. The age of machinery. In Mingay (1981), I, 200–13.

Collins, E.J.T. 1987. The rationality of surplus agricultural labour: mechanisation in English agriculture in the nineteenth century, *AHR*, 35: 36–46.

Collins, K. 1967. Marx on the English agricultural revolution: theory and evidence, *History and Theory*, 6: 351–81.

Colyer, R.J. 1975. The use of estate home farm accounts as sources for nineteenth century agricultural history, *Local Historian*, 11: 406–13.

Cooper, J.P. 1967. The social distribution of land and men in England, 1436–1700, *EcHR*, 20: 419–40. Reprinted with postscript in Floud (1974), 107–32.

Coppock, J.T. 1984. Mapping the agricultural returns: a neglected tool of historical geography. In Reed (1984b), 167–94.

Copus, A.K. 1989. Changing markets and the development of sheep breeds in southern England, 1750–1900, *AHR*, 37: 36–51.

Cornwall, J. 1954. Farming in Sussex, 1560–1640, *Sussex Archaeological Collections*, 92: 48–92.

Cornwall, J. 1960. Agricultural improvement, 1560–1640, *Sussex Archaeological Collections*, 98: 118–32.

Cornwall, J.C.K. 1988. *Wealth and society in early sixteenth century England*.

Cosgrove, D. and Daniels, S. (eds.) 1988. *The iconography of landscape: essays on the symbolic representation, design and use of past environments*, Cambridge.

Coward, B. 1988. *Social change and continuity in early modern England 1550–1750.*

Crafts, N.F.R. 1977. Determinants of the rate of parliamentary enclosure, *EEcH*, 14: 227–49.

Crafts, N.F.R. 1983. British economic growth 1700–1831: a review of the evidence, *EcHR*, 36: 177–99.

Crafts, N.F.R. 1985. *British economic growth during the Industrial Revolution,* Oxford.

Crafts, N.F.R. 1987. British economic growth, 1700–1850: some difficulties of interpretation, *EEcH*, 24: 245–68.

Craigie, P.G. 1883. Statistics of agricultural production, *Journal of the Royal Statistical Society,* 46: 1–58.

Creasey, J. 1981. Agrarian and food history. In Lilley (1981).

Cressy, D. 1986. Kinship and kin interaction in early modern England, *P&P*, 113: 38–69.

Curtler, W.H.R. 1909. *A short history of English agriculture,* Oxford.

Curtler, W.H.R. 1920. *The enclosure and redistribution of our land,* Oxford.

Dahlman, C.J. 1980. *The open field system and beyond,* Cambridge.

Darby, H.C. 1951. The changing English landscape, *Geographical Journal,* 117: 377–94.

Darby, H.C. 1954. Some early ideas on the agricultural regions of England, *AHR*, 2: 30–47.

Darby, H.C. 1964. The draining of the English clay-lands, *Geographische Zeitschrift,* 52: 190–201.

Darby, H.C. 1973a. The age of the improver: 1600–1800. In Darby (1973b), 302–87.

Darby, H.C. 1983. *The changing fenland,* Cambridge.

Darby, H.C. (ed.) 1936. *An historical geography of England before A.D. 1800,* Cambridge.

Darby, H.C. (ed.) 1973b. *A new historical geography of England,* Cambridge.

Davie, N. 1991. Chalk and cheese? 'Fielden' and 'Forest' communities in early modern England, *Journal of Historical Sociology,* 4: 1–31.

Davis, J. and Mathias, P. (eds.) 1995. *The nature of industrialization,* IV, *Agriculture and industrialization,* Oxford.

Deane, P. and Cole, W.A. 1967. *British economic growth, 1688–1959,* 2nd edn, Cambridge.

Devine, T.M. 1994. *The transformation of rural Scotland: social change and the agrarian economy, 1660–1815,* Edinburgh.

Digby, A. and Feinstein, C.H. (eds.) 1989. *New directions in economic and social history.*

Dodd, P. 1987. The agricultural statistics for 1854: an assessment of their value, *AHR*, 35: 159–70.

Dodgshon, R.A. 1980. *The origin of British field systems: an interpretation.*

Dodgshon, R.A. 1987. *The European past: social evolution and spatial order.*

Dodgshon, R.A. 1990. The changing evolution of space 1500–1914. In Dodgshon and Butlin (1990), 255–84.

Dodgshon, R.A. and Butlin, R.A. (eds.) 1990. *An historical geography of England and Wales,* 2nd edn.

Drury, J.L. 1987. More stout than wise: tenant right in Weardale in the Tudor period. In Marcombe (1987), 71–100.

Dyer, C. 1992. The hidden trade of the middle ages: evidence from the west Midlands of England, *JHG*, 18: 141–57.

Dyer, C. 1995. Were peasants self sufficient: English villagers and the market, 900–1350. In Mornet (1995), 653–66.

Edwards, P. 1978. The development of dairy farming on the north Shropshire plain in the seventeenth century, *Midland History*, 4: 175–89.

Edwards, P. 1981. The cattle trade of Shropshire in the late sixteenth and seventeenth centuries, *Midland History*, 6: 72–94.

Edwards, P. 1988. *The horse trade of Tudor and Stuart England*, Cambridge.

Edwards, P. 1991. *Farming: sources for local historians.*

Ellis, H. 1840. Speculi Britanniae pars: *an historical and chorographical description of the county of Essex, by John Norden, 1594*, Camden Society, old series.

Emery, F.V. 1974. *The Oxfordshire landscape.*

Emery, F.V. 1976. The mechanics of innovation: clover cultivation in Wales before 1750, *JHG*, 2: 35–48.

English, B. and Saville, J. 1983. *Strict settlement: a guide for historians*, University of Hull, Occasional Papers in Economic and Social History 10, Hull.

Ernle, Lord, 1961. *English farming past and present*, 6th edn.

Evans, E. 1976. *The contentious tithe: the tithe problem and English agriculture, 1750–1850.*

Evans, G.E. 1960. *The horse in the furrow.*

Evans, G.E. 1966. *The pattern under the plough.*

Evans, G.E. 1969. *The farm and the village.*

Evans, N. 1984. Farming and land-holding in wood-pasture East Anglia 1550–1650, *Proceedings of the Suffolk Institute for Archaeology and Natural History*, 35: 303–15.

Everitt, A. 1966. *The community of Kent and the Great Rebellion*, Leicester.

Everitt, A. 1967a. Farm labourers. In Thirsk (1967d), 396–465.

Everitt, A. 1967b. The marketing of agricultural produce. In Thirsk (1967d), 466–592.

Everitt, A. 1970. Nonconformity in country parishes. In Thirsk (1970b), 178–99.

Everitt, A. 1979. Country, county and town: problems of regional evolution in England, *Transactions of the Royal Historical Society*, 5th series, 29: 79–108. Reprinted in Everitt (1985), 11–40.

Everitt, A. 1985. *Landscape and community in England.*

Fairlie, S. 1969. The corn laws and British wheat production, *EcHR*, 22: 88–116.

Feinstein, C.H. 1978. Capital formation in Great Britain. In Mathias and Postan (1978), 28–96.

Feinstein, C.H. 1988. Agriculture. In Feinstein and Pollard (1988), 267–80.

Feinstein, C.H. and Pollard, S. (eds.) 1988. *Studies in capital formation in the United Kingdom, 1750–1920*, Oxford.

Fenoaltea, S. 1988. Transaction costs, Whig history, and the common fields, *Politics and Society*, 16: 171–240.

Fieldhouse, R. 1980. Agriculture in Wensleydale since 1600, *Northern History*, 16: 67–95.

Fisher, F.J. 1935. The development of the London food market, 1540–1640, *Economic History Review*, 5: 46–64. Reprinted in Carus-Wilson (1954–62), I, 135–51.

Fisher, F.J. (ed.) 1961. *Essays in the economic and social history of Tudor and Stuart England in honour of R. H. Tawney*, Cambridge.

Fisher, J.A. 1983. Landowners and English tenant right, 1845–1852, *AHR*, 31: 15–25.

Fletcher, A. and Stevenson, J. (eds.) 1985. *Order and disorder in early modern England*, Cambridge.

Flinn, M.W. 1966. Agricultural productivity and economic growth: a comment, *JEcH*, 26: 93–8.

Floud, R. (ed.) 1974. *Essays in quantitative economic history*, Oxford.

Floud, R. and McCloskey, D. (eds.) 1981. *The economic history of Britain since 1700*, 2 vols., Cambridge.

Floud, R. and McCloskey, D. (eds.) 1994. *The economic history of Britain since 1700*, 2nd edn, 2 vols., Cambridge.

Fox, H.S.A. 1979. Local farmers' associations and the circulation of agricultural information in nineteenth-century England. In Fox and Butlin (1979), 43–63.

Fox, H.S.A. and Butlin, R.A. (eds.) 1979. *Change in the countryside: essays on rural England 1500–1900*, Institute of British Geographers Special Publication 10.

Fream, W. 1892. *Elements of agriculture*.

Frearson, M. 1994. Communications and the continuity of dissent in the Chiltern Hundreds during the sixteenth and seventeenth centuries. In Spufford (1994), 288–308.

Frost, P. 1981. Yeomen and metalsmiths: livestock in the dual economy in south Staffordshire 1560–1720, *AHR*, 29: 29–41.

Fussell, G.E. 1929. The size of English cattle in the eighteenth century, *AH*, 3: 160–81.

Fussell, G.E. 1955. Adventures with clover, *Agriculture*, 52: 342–5.

Fussell, G.E. 1959. The Low Countries influence on English farming, *English Historical Review*, 74: 611–22.

Fussell, G.E. 1964. 'Norfolk improvers': their farms and methods, *Norfolk Archaeology*, 33: 332–44.

Fussell, G.E. 1965. *Farming techniques from prehistoric to modern times*, Oxford.

Fussell, G.E. 1973. *Jethro Tull: his influence on mechanised agriculture*, Reading.

Fussell, G.E. 1984. *Landscape painting and the agricultural revolution*.

Fussell, G.E. (ed.) 1936. *Robert Loder's farm accounts 1610–1620*, Camden Society, 3rd series, 53.

Fussell, G.E. and Goodman, C. 1936. Eighteenth-century traffic in live-stock, *Economic History*, 3: 214–36.

Galenson, D.W. (ed.) 1989. *Markets in history: economic studies of the past*, Cambridge.

Garnier, R.M. 1896. The introduction of forage crops into Great Britain, *JRASE*, 3rd series, 7: 82–97.

Gazley, J.G. 1973. *The life of Arthur Young, 1741–1820*, Philadelphia.

Gibson, A.J.S. 1988. The size and weight of cattle and sheep in early modern Scotland, *AHR*, 36: 162–71.

Giles, B.D. 1956. Agriculture and the price mechanism. In Wilson and Andrews (1956), 173–203.

Glass, D.V. and Eversley, D.E.C. (eds.) 1965. *Population in history*.

Glennie, P. 1988a. Continuity and change in Hertfordshire agriculture, 1550–1700: I – patterns of agricultural production, *AHR*, 36: 55–76.

Glennie, P. 1988b. Continuity and change in Hertfordshire agriculture, 1550–1700: II – trends in crop yields and their determinants, *AHR*, 36: 145–61.

Glennie, P. 1988c. In search of agrarian capitalism: manorial land markets and the acquisition of land in the Lea Valley, *c*.1450–*c*.1560, *Continuity and Change*, 3: 11–40.

Glennie, P. 1991. Measuring crop yields in early modern England. In Campbell and Overton (1991), 255–83.

Goddard, N. 1988. *Harvests of change: the Royal Agricultural Society of England 1838–1988*.

Goddard, N. 1989. Agricultural literature and societies. In Mingay (1989), 361–83.

Goddard, N. 1991. Information and innovation in early-Victorian farming systems. In Holderness and Turner (1991), 165–90.

Goldstone, J. 1988. Regional ecology and agrarian development in England and France, *Politics and Society*, 16: 287–334.

Gonner, E.C.K. 1912. *Common land and inclosure*.

Goodman, D. and Redclift, M. 1981. *From peasant to proletarian: capitalist development and agrarian transitions*, Oxford.

Grantham, G. 1989a. Agricultural supply during the industrial revolution: French evidence and European implications, *JEH*, 49: 43–72.

Grantham, G. 1989b. Jean Meuvret and the subsistence problem in early modern France, *JEH*, 49: 184–200.

Grantham, G. and Leonard, C.S. (eds.) 1989. *Agrarian organization in the century of industrialization: Europe, Russia, and North America*, Research in Economic History, Supplement 5, Greenwich, Conn.

Gras, N.S.B. 1915. *The evolution of the English corn market from the twelfth to the eighteenth century*, Cambridge, Mass.

Gras, N.S.B. 1925. *A history of agriculture*, New York.

Gregson, N. 1989. Tawney revisited: custom and the emergence of capitalist class relations in north east Cumbria 1600–1830, *EcHR*, 42: 18–42.

Grigg, D.B. 1963. Small and large farms in England and Wales: their size and distribution, *Geography*, 48: 268–79.

Grigg, D.B. 1966. *The agricultural revolution in south Lincolnshire*, Cambridge.

Grigg, D.B. 1967a. Regions, models and classes. In Chorley and Haggett (1967), 461–507.

Grigg, D.B. 1967b. The changing agricultural geography of England: a commentary on the sources available for the reconstruction of the agricultural geography of England, 1770–1850, *TIBG*, 41: 73–96.

Grigg, D.B. 1979. Ester Boserup's theory of agrarian change: a critical review, *Progress in Human Geography*, 3: 64–84.

Grigg, D.B. 1980. *Population growth and agrarian change: an historical perspective*, Cambridge.

Grigg, D.B. 1982. *The dynamics of agricultural change*.

Habakkuk, H.J. 1940. English landownership 1680–1740, *EcHR*, 10: 2–17.

Habakkuk, H.J. 1965. Landowners and the Civil War, *EcHR*, 18: 130–51.

Habakkuk, H.J. 1979–81. The rise and fall of English landed families, 1660–1800, *Transactions of the Royal Historical Society*, 29 (1979): 187–207; 30 (1980): 199–221; 31 (1981): 195–217.

Habakkuk, H.J. 1987. The agrarian history of England and Wales: regional farming systems and agrarian change, 1640–1750, *EcHR*, 40: 281–96.

Habakkuk, H.J. 1994. *Marriage, debt, and the estates system: English landownership 1650–1950*, Oxford.

Hagen, W.W. 1988. Capitalism in the countryside in early modern Europe: interpretations, models, debates, *AH*, 62: 13–47.

Hainsworth, D.R. 1992. *Stewards, lords and people: the estate steward and his world in later Stuart England*, Cambridge.

Hall, J.G. and Clutton-Brock, J. 1989. *Two hundred years of British farm livestock*.

Hamer, J.H. 1968. Trading at Saint White Down Fair, 1637–1649, *Somerset Archaeological and Natural History Society*, 112: 61–70.

Hammond, J.L. and Hammond, B. 1912. *The village labourer*.

Harris, A. 1961. *The rural landscape of the East Riding of Yorkshire, 1700–1850*.

Harte, N.B. and Ponting, K.G. (eds.) 1973. *Textile history and economic history: essays in honour of Miss Julia de Lacy Mann*, Manchester.

Hartlib, S. 1651. *His legacie, or an enlargement of the husbandry used in Brabant and Flanders*.

Harvey, N. 1949. The coming of the swede to Great Britain: an obscure chapter in farming history, *AH*, 23: 286–8.

Havinden, M.A. 1961. Agricultural progress in open-field Oxfordshire, *AHR*, 9: 73–83. Reprinted in Jones (1967a), 66–79 and in Minchinton (1968), I, 147–59.

Havinden, M.A. 1965. *Household and farm inventories in Oxfordshire, 1550–90*, Oxford Record Society 44 and Historical Manuscripts Commission Joint Publication 10.

Havinden, M.A. 1974. Lime as a means of agricultural improvement: the Devon example. In Chalklin and Havinden (1974), 104–34.

Heal, F. and Holmes, C. 1994. *The gentry in England and Wales, 1500–1700*, Basingstoke.

Hey, D.G. 1969. A dual economy in south Yorkshire, *AHR*, 17: 108–19.

Hey, D.G. 1974. *An English rural community, Myddle under the Tudors and Stuarts*, Leicester.

Hey, D.G. 1980. *Packmen, carriers and packhorse roads*, Leicester.

Hidden, N. 1989. Jethro Tull I, II, and III, *AHR*, 37: 26–35.

Hill, B. 1989. *Women, work, and sexual politics in eighteenth-century England*, Oxford.

Hill, C. 1961. *The century of revolution*.

Hill, C. 1969. *Reformation to revolution*.

Hilton, R.H. (ed.) 1976. *The transition from feudalism to capitalism*.

Historical Manuscripts Commission 1907. *Report on manuscripts in various collections*, IV.

Historical Manuscripts Commission 1911. *Report of the manuscripts of Lord Middleton, preserved at Wollaton Hall, Nottinghamshire*.

Hobsbawm, E.J. and Rudé, G. 1969. *Captain Swing*.

Hodges, R. 1988. *Primitive and peasant markets*, Oxford.

Hodgshon, R.I. 1979. The progess of enclosure in County Durham. In Fox and Butlin (1979), 83–102.

Holderness, B.A. 1974. The English land market in the eighteenth century: the case of Lincolnshire, *EcHR*, 27: 557–76.

Holderness, B.A. 1976a. Credit in English rural society before the nineteenth century, with special reference to the period 1650–1720, *AHR*, 24: 97–109.

Holderness, B.A. 1976b. *Pre-industrial England: economy and society 1500–1750*.

Holderness, B.A. 1981. The Victorian farmer. In Mingay (1981), I, 227–44.

Holderness, B.A. 1988. Agriculture, 1770–1860. In Feinstein and Pollard (1988), 9–34.

Holderness, B.A. 1989. Prices, productivity and output. In Mingay (1989), 84–189.

Holderness, B.A. 1991. The origins of high farming. In Holderness and Turner (1991), 149–64.

Holderness, B.A. and Turner, M. 1991. *Land, labour and agriculture, 1700–1920: essays for Gordon Mingay*.

Holton, R.J. 1985. *The transition from feudalism to capitalism*.

Hopcroft, R.L. 1994. The social origins of agrarian change in late medieval England, *American Journal of Sociology*, 6: 1559–95.

Horn, P. 1982. *William Marshall (1745–1818) and the Georgian countryside*, Sutton Courtney.

Hoskins, W.G. 1943. The reclamation of waste in Devon 1550–1800, *Economic History Review*, 12: 80–92.

Hoskins, W.G. 1945. The Leicestershire farmer in the sixteenth century, *Transactions of the Leicestershire Archaeological Society*, 22: 33–95. Revised version in Hoskins (1950), 123–83.

Hoskins, W.G. 1949. *Midland England*.

Hoskins, W.G. 1951. The Leicestershire farmer in the seventeenth century, *AH*, 25: 9–20. Reprinted in Hoskins (1963), 149–69.

Hoskins, W.G. 1954. Regional farming in England, *AHR*, 2: 3–11.

Hoskins, W.G. 1955. *The making of the English landscape*.

Hoskins, W.G. 1957. *The Midland peasant*.

Hoskins, W.G. 1963. *Provincial England*.

Hoskins, W.G. 1964. Harvest fluctuations and English economic history, 1480–1619, *AHR*, 12: 28–46. Reprinted in Minchinton (1968), I, 93–116.

Hoskins, W.G. 1968. Harvest fluctuations and English economic history, 1620–1759, *AHR*, 16: 15–31.

Hoskins, W.G. 1976. *The age of plunder*.

Hoskins, W.G. (ed.) 1948. *Studies in Leicestershire agrarian history*, Leicester.

Hoskins, W.G. (ed.) 1950. *Essays in Leicestershire history*, Leicester.

Houghton, J. 1692–1703. *A collection for the improvement of husbandry and trade*, 9 vols., ed. R. Bradley, in 4 vols., 1727–8. Reprinted Farnborough, 1969.

Hoyle, R.W. 1987. An ancient and laudable custom: the definition and development of tenant right in north-western England in the sixteenth century, *P&P*, 116: 24–55.

Hoyle, R.W. 1990. Tenure and the land market in early modern England: or a late contribution to the Brenner debate, *EcHR*, 43: 1–20.

Hoyle, R. 1992a. Disafforestation and drainage: the Crown as entrepreneur. In Hoyle (1992b), 353–88.

Hoyle, R. (ed.) 1992b. *The estates of the English Crown, 1558–1640*, Cambridge.

Hudson, K. 1972. *Patriotism with profit: British agricultural societies in the eighteenth and nineteenth centuries.*

Hueckel, G. 1976. Relative prices and supply response in English agriculture during the Napoleonic Wars, *EcHR*, 29: 401–14.

Hueckel, G. 1981. Agriculture during industrialisation. In Floud and McCloskey (1981), I, 182–203.

Humphries, J. 1990. Enclosures, common rights, and women: the proletarianization of families in the late eighteenth and early nineteenth centuries, *JEcH*, 50: 17–42.

Hunt, E.H. 1967. Labour productivity in English agriculture, 1850–1914, *EcHR*, 20: 280–92.

Hunt, E.H. 1986. Wages. In Langton and Morris (1986), 60–8.

Hutton, R. 1994. *The rise and fall of merry England: the ritual year 1400–1700*, Oxford.

Ingram, M. 1987. *Church courts, sex and marriage in England, 1570–1640*, Cambridge.

Jackson, R.V. 1985. Growth and deceleration in English agriculture 1660–1790, *EcHR*, 38: 333–51.

Jenkins, H.M. 1869. The Lodge Farm, Castle Acre, Norfolk, in the occupation of Mr John Hudson, *JRASE*, 2nd series, 5: 460–74.

Jewell, C.A. 1965. *Victorian farming: a sourcebook*, Winchester.

John, A.H. 1960. The course of agricultural change, 1660–1760. In Pressnell (1960), 125–55. Reprinted in Minchinton (1968), I, 223–53.

John, A.H. 1961. Aspects of economic growth in the first half of the eighteenth century, *Economica*, new series, 28: 176–90. Reprinted in Minchinton (1969), 165–83, and in Carus-Wilson (1954–62), II, 360–73.

John, A.H. 1965. Agricultural productivity and economic growth in England, 1700–1760, *JEcH*, 25: 19–34. Reprinted in Jones (1967b), 172–93.

John, A.H. 1976. English agricultural improvement and grain exports, 1660–1765. In Coleman and John (1976), 47–51.

John, A.H. 1989. Statistical appendix. In Mingay (1989), 972–1155.

Jones, E.L. 1960. Eighteenth-century changes in Hampshire chalkland farming, *AHR*, 8: 5–19. Reprinted in Jones (1974), 23–40.

Jones, E.L. 1962. The changing basis of English agricultural prosperity, 1853–1873, *AHR*, 10: 102–19. Reprinted in Jones (1974), 191–210.

Jones, E.L. 1964a. *Seasons and prices: the role of the weather in English Agricultural History.*

Jones, E.L. 1964b. The agricultural labour market in England 1793–1872, *EcHR*, 17: 322–38. Reprinted in Jones (1974), 211–30.

Jones, E.L. 1965. Agriculture and economic growth in England, 1660–1750: agricultural change, *JEcH*, 25: 1–18. Reprinted in Jones (1967b), 152–71; in Jones (1974), 67–84; and in Minchinton (1968), I, 205–19.

Jones, E.L. 1967a. Editor's introduction. In Jones (1967b), 1–48. Reprinted in

Jones (1974), 85–127.

Jones, E.L. 1968. The condition of English agriculture 1500–1640, *EcHR*, 21: 614–19.

Jones, E.L. 1974. *Agriculture and the industrial revolution*, Oxford.

Jones, E.L. 1975. Afterword. In Parker and Jones (1975), 327–60.

Jones, E.L. 1981. Agriculture 1700–80. In Floud and McCloskey (1981), 66–86.

Jones, E.L. (ed.) 1967b. *Agriculture and economic growth in England 1650–1815.*

Jones, E.L. and Woolf, S.J. (eds.) 1969. *Agrarian change and economic development.*

Kain, R.J.P. 1986. *An atlas and index of the tithe files of mid-nineteenth-century England and Wales*, Cambridge.

Kain, R.J.P. and Prince, H. C. 1985. *The tithe surveys of England and Wales*, Cambridge.

Kent, N. 1796. *General view of the agriculture of the county of Norfolk.*

Kenyon, G.H. 1955. Kirdford inventories, 1611–1776, with particular reference to the weald clay farming, *Sussex Archaeological Collections*, 93: 78–157.

Kerridge, E. 1953. The sheep fold in Wiltshire and the floating of watermeadows, *EcHR*, 6: 282–9.

Kerridge, E. 1959. Agriculture 1500–1793. In *The Victoria histories of the counties of England. A history of Wiltshire,* IV, 43–64.

Kerridge, E. 1967. *The agricultural revolution.*

Kerridge, E. 1968. Arthur Young and William Marshall, *History Studies*, 1: 43–53.

Kerridge, E. 1969a. *Agrarian problems in the sixteenth century and after.*

Kerridge, E. 1969b. The agricultural revolution reconsidered, *AH*, 43: 463–76.

Kerridge, E. 1973. *The farmers of old England.*

Kerridge, E. 1986. Early modern English markets. In Anderson and Latham (1986), 121–54.

Kerridge, E. 1988. *Trade and banking in early modern England*, Manchester.

Kerridge, E. 1992. *The common fields of England*, Manchester.

King, P. 1989. Gleaners, farmers and the failure of legal sanctions in England 1750–1850, *P&P*, 125: 116–150.

King, P. 1991. Customary rights and women's earnings: the importance of gleaning to the rural labouring poor, 1750–1850, *EcHR*, 44: 461–76.

Kussmaul, A. 1981. *Servants in husbandry in early modern England*, Cambridge.

Kussmaul, A. 1990. *A general view of the rural economy of England, 1538–1840*, Cambridge.

Landau, N. 1988. The laws of settlement and the surveillance of immigration in eighteenth-century Kent, *Continuity and Change*, 3: 391–420.

Lane, C. 1980. The development of pastures and meadows during the sixteenth and seventeenth centuries, *AHR*, 28: 18–30.

Langdon, J.L. 1986. *Horses, oxen and technological innovation: the use of draught animals in English farming from 1066–1500*, Cambridge.

Langton, J. and Morris, R. (eds.) 1986. *An atlas of industrializing Britain 1780–1914.*

Large, P. 1984. Urban growth and agricultural change in the West Midlands during the seventeenth and eighteenth centuries. In Clark (1984), 169–89.

Laslett, P. 1983. *The world we have lost further explored.*

Lavrovsky, V.M. 1956. Expropriation of the English peasantry in the eighteenth century, *EcHR*, 9: 271–82.

Lazonick, W. 1974. Karl Marx and enclosures in England, *Review of Radical Political Economy*, 6: 1–59.

Lennard, R.V. 1932. English agriculture under Charles II, *Economic History Review*, 4: 23–45.

Lennard, R.V. 1964. Agrarian history: some vistas and pitfalls, *AHR*, 12: 83–98.

Levine, D. and Wrightson, K. 1991. *The making of an industrial society: Wickham, 1560–1765*, Oxford.

Lilley, G.P. (ed.) 1981. *Information sources in agriculture and food science.*

Lindley, K. 1982. *Fenland riots and the English revolution.*

Lipton, M. 1990. Responses to rural population growth: Malthus and the moderns. In McNicoll and Cain (1990), 215–42.

Lisle, E. 1757. *Observations in husbandry*, 2nd edn, 2 vols. Reprinted, Farnborough, 1970.

Lodge, E.C. (ed.) 1927. *The account book of a Kentish estate 1616–1704*, Records of the Social and Economic History of England and Wales, 1st series.

Long, W.H. 1960. Regional farming in seventeenth-century Yorkshire, *AHR*, 8: 103–14.

Longman, G. 1977. *A corner of England's garden: an agrarian history of south west Hertfordshire, 1600–1850*, 2 vols.

Lucas, J. (trans.) 1892. *Kalm's account of his visit to England on his way to America in 1748.*

McCloskey, D.N. 1972. The enclosure of the open fields: preface to a study of its impact on the efficiency of English agriculture in the eighteenth century, *JEcH*, 32: 15–35.

McCloskey, D.N. 1975a. The economics of enclosure: a market analysis. In Parker and Jones (1975), 123–60.

McCloskey, D.N. 1975b. The persistence of English common fields. In Parker and Jones (1975), 73–119.

McCloskey, D.N. 1981. The industrial revolution 1780–1860: a survey. In Floud and McCloskey (1981), I, 103–27.

McCloskey, D.N. 1989. The open fields of England: rent, risk, and the rate of interest, 1300–1815. In Galenson (1989), 5–51.

MacCulloch, D. 1979. Kett's rebellion in context, *P&P*, 84: 36–59.

Macdonald, S. 1975. The progress of the early threshing machine, *AHR*, 23: 63–77.

Macdonald, S. 1976. The lease in agricultural improvement, *JRASE*, 147: 19–26.

Macdonald, S. 1980. Agricultural response to a changing market during the Napoleonic Wars, *EcHR*, 33: 59–71.

Macfarlane, A. 1976. *The diary of Ralph Josselin, 1616–1683*, Records of Social and Economic History, new series, 3.

Macfarlane, A. 1977. History, anthropology and the study of communities, *Social History*, 5: 631–52.

Macfarlane, A. 1978. *The origins of English individualism*, Oxford.

Macfarlane, A. 1987. *The culture of capitalism*, Oxford.

McGregor, O.R. 1961. Introduction Part two: English farming after 1815. In Ernle (1961), lxxix–cxlv.

McNicoll, G. and Cain, M. (eds.) 1990. *Rural development and population: institutions and policy*, supplement to volume 15 of *Population and Development Review*, Oxford.

McQuiston, J.R. 1973. Tenant right: farmer against landlord in Victorian England 1847–1883, *AH*, 47: 95–113.

Malcolmson, R.W. 1981. *Life and labour in England, 1700–1780.*

Manning, R.B. 1988. *Village revolts: social protest and popular disturbances in England 1509–1640*, Oxford.

Marcombe, D. (ed.) 1987. *The last principality: politics, religion and society in the Bishopric of Durham, 1494–1660*, Nottingham.

Marshall, G. 1978. The Rotheram plough: a study of a novel 18th century implement of agriculture, *Tools and Tillage*, 3: 149–67.

Marshall, J. 1980. Agrarian wealth and social structure in pre-industrial Cumbria, *EcHR*, 33: 503–21.

Marshall, J.D. 1985. *The old poor law 1795–1834*, Studies in Economic and Social History, Basingstoke.

Marshall, W. 1778. *Minutes of agriculture made on a farm of 300 acres of various soils, near Croydon, Surrey.*

Marshall, W. 1779. *Experiments and observations concerning agriculture and the weather.*

Marshall, W. 1787. *The rural economy of Norfolk*, 2 vols.

Martin, J.E. 1983. *Feudalism to capitalism: peasant and landlord in English agrarian development.*

Martin, J.E. 1988. Sheep and enclosure in sixteenth-century Northamptonshire, *AHR*, 36: 39–45.

Mathew, W.M. 1993. Marling in British agriculture: a case of partial identity, *AHR*, 41: 97–110.

Mathias, P. and Postan, M. (eds.) 1978. *The Cambridge economic history of Europe*,VII, *The industrial economies: capital, labour, and enterprise*, Cambridge.

Medick, H. 1976. The proto-industrial family economy: the structural function of household and family during the transition from peasant to industrial capitalism, *Social History*, 3: 291–15.

Medick, H. 1981. The transition from feudalism to capitalism: renewal of the debate. In Samuel (1981), 120–30.

Mendels, F.F. 1972. Proto-industrialization: the first phase of the industrialization process, *JEcH*, 32: 241–61.

Mills, D. (ed.) 1973. *English rural communities: the impact of a specialised economy.*

Minchinton, W.E. 1953. Agricultural returns and the Government during the Napoleonic Wars, *AHR*, 1: 29–43. Reprinted in Minchinton (1968), II, 103–20.

Minchinton, W.E. 1971–2. The agricultural regions of England and Wales, *Proceedings of the Hungarian Agricultural Museum*, 109–20.

Minchinton, W.E. (ed.) 1968. *Essays in agrarian history*, 2 vols., Newton Abbot.

Minchinton, W.E. (ed.) 1969. *The growth of English overseas trade in the sixteenth and seventeenth centuries.*

Minchinton, W.E. (ed.) 1981. *University of Exeter Papers in Economic History*, 14, Exeter.

Mingay, G.E. 1956. The agricultural depression, 1730–1750, *EcHR*, 8: 323–38. Reprinted in Carus-Wilson (1954–62), II, 309–26.

Mingay, G.E. 1962. The size of farms in the eighteenth century, *EcHR*, 14: 469–88.

Mingay, G.E. 1963a. *English landed society in the eighteenth century.*

Mingay, G.E. 1963b. The 'agricultural revolution' in English history: a reconsideration, *AH*, 37: 123–33. Reprinted in Minchinton (1968), II, 9–28.

Mingay, G.E. 1968. *Enclosure and the small farmer in the age of the industrial revolution*, Studies in Economic and Social History, Basingstoke.

Mingay, G.E. 1969. Dr. Kerridge's 'agricultural revolution'; a comment, *AH*, 43: 477–81.

Mingay, G.E. 1975. *Arthur Young and his times.*

Mingay, G.E. 1976. *The gentry: the rise and fall of a ruling class.*

Mingay, G.E. (ed.) 1981. *The Victorian countryside*, 2 vols.

Mingay, G.E. (ed.) 1989. *The agrarian history of England and Wales*, VI, *1750–1850*, Cambridge.

Mitchell, B.R. and Deane, P. 1962. *Abstract of British historical statistics*, Cambridge.

Mitchison, R. 1959. The Old Board of Agriculture, *English Historical Review*, 74: 41–69.

Mokyr, J. 1987. Has the industrial revolution been crowded out? Some reflections on Crafts and Williamson, *EEcH*, 24: 293–391.

Mokyr, J. (ed.) 1985. *The economics of the industrial revolution.*

Mokyr, J. (ed.) 1993. *The British industrial revolution: an economic perspective*, Oxford.

Moore, H.I. 1946. *Grassland husbandry*, 3rd edn.

Moore, J.S. 1977. *The goods and chattels of our forefathers: Frampton Cotterell and district probate inventories, 1539–1804*, Chichester.

Morgan, D.H. 1975. The place of harvesters in nineteenth-century village life. In Samuel (1975), 27–72.

Morgan, R. 1984. *Farm tools, implements and machines in Britain, pre-history to 1945: a bibliography*, Reading.

Morgan, R. 1989. Root crops. In Mingay (1989), 296–304.

Mornet, E. 1995. *Campagnés médiévales: l'homme et son espace: études offertes a Robert Fossier*, Paris.

Morrill, J. 1987. The ecology of allegiance in the English revolution, *Journal of British Studies*, 26: 451–67. Reprinted as, The ecology of allegiance in the English Civil Wars, in Morrill (1993), 224–41.

Morrill, J. 1993. *The nature of the English revolution.*

Munsche, P.B. 1981. *Gentlemen and poachers: the English game laws 1671–1831*, Cambridge.

Nair, G. 1988. *Highley: the development of a community, 1550–1880*, Oxford.

Neeson, J.M. 1984. The opponents of enclosure in eighteenth-century Northamptonshire, *P&P*, 105: 114–39.

Neeson, J.M. 1989. Parliamentary enclosure and the disappearance of the English peasantry, revisited. In Grantham and Leonard (1989), 89–120.

Neeson, J.M. 1993. *Commoners: common right, enclosure and social change in England, 1700–1820*, Cambridge.

O'Brien, P. 1977. Agriculture and the industrial revolution, *EcHR*, 30: 166–81.

O'Brien, P. 1985. Agriculture and the home market for English industry, 1660–1820, *English Historical Review*, 50: 773–800.

Oddy, D. and Miller, D. (eds.) 1976. *The making of the modern British diet.*

Ormrod, D. 1985. *English grain exports and the structure of agrarian capitalism 1700–1760*, University of Hull Occasional Paper in Economic and Social History.

Orwin, C.S. 1949. *A history of English farming.*

Orwin, C.S. and Orwin, C.S. 1938. *The open fields.*

Orwin, C.S. and Whetham, E. 1964. *The history of British agriculture 1846–1914.*

Outhwaite, R.B. 1969. *Inflation in Tudor and early Stuart England.* Studies in Economic and Social History, Basingstoke.

Outhwaite, R.B. 1981. Dearth and government intervention in English grain markets, 1590–1700, *EcHR*, 33: 389–406.

Outhwaite, R.B. 1991. *Dearth, public policy and social disturbance in England, 1550–1800*, Studies in Economic and Social History, Basingstoke.

Overton, M. 1979. Estimating crop yields from probate inventories: an example from East Anglia, 1585–1735, *JEcH*, 39: 363–78.

Overton, M. 1983a. *Agricultural regions in early modern England: an example from East Anglia*, University of Newcastle upon Tyne, Department of Geography Seminar Paper 43.

Overton, M. 1983b. An agricultural revolution, 1650–1750. In Overton *et al.* (1983), 9–13.

Overton, M. 1984a. Agricultural productivity in eighteenth-century England: some further speculations, *EcHR*, 37: 244–51.

Overton, M. 1984b. Agricultural revolution? Development of the agrarian economy in early modern England. In Baker and Gregory (1984), 118–39.

Overton, M. 1984c. Probate inventories and the reconstruction of agrarian landscapes. In Reed (1984b), 167–94.

Overton, M. 1985. The diffusion of agricultural innovations in early modern England: turnips and clover in Norfolk and Suffolk 1580–1740, *TIBG*, new series, 10: 205–21.

Overton, M. 1986a. Agriculture. In Langton and Morris (1986), 34–53.

Overton, M. 1986b. Depression or revolution? English agriculture 1640–1750, *Journal of British Studies*, 25: 344–52.

Overton, M. 1988. Agrarian history. In Cannon (1988), 5–7.

Overton, M. 1989a. Agricultural revolution? England, 1540–1850. In Digby and Feinstein (1989), 9–21.

Overton, M. 1989b. Weather and agricultural change in England, 1660–1739, *AH*, 43: 77–88.

Overton, M. 1990a. Re-estimating crop yields from probate inventories, *JEcH*, 50: 931–5.

Overton, M. 1990b. The critical century? The agrarian history of England and Wales 1750–1850, *AHR*, 38: 185–9.

Overton, M. 1991. The determinants of crop yields in early modern England. In Campbell and Overton (1991), 284–322.

Overton, M. 1995. Land and labour productivity in English agriculture, 1650–1850. In Davis and Mathias (1995).

Overton, M. 1996. Re-establishing the agricultural revolution, *AHR*, 44.

Overton, M. and Campbell, B.M.S. 1991. Productivity change in European agricultural development. In Campbell and Overton (1991), 1–50.

Overton, M. and Campbell, B.M.S. 1992. Norfolk livestock farming 1250–1740: a comparative study of manorial accounts and probate inventories, *JHG*, 18: 377–96.

Overton, M. *et al.* 1983. *Agricultural history: papers presented to the Economic History Society Conference*, Canterbury.

Palliser, D.M. 1976. *The Staffordshire landscape.*

Palliser, D.M. 1982. Tawney's century: brave new world or Malthusian trap?, *EcHR*, 35: 339–54.

Palliser, D.M. 1983. *The age of Elizabeth: England under the later Tudors 1547–1603.*

Parker, L.A. 1948. The agrarian revolution at Cotesbach, 1501–1612. In Hoskins (1948), 41–76.

Parker, R.A.C. 1975. *Coke of Norfolk: a financial and agricultural study, 1707–1842*, Oxford.

Parker, W.N. and Jones, E.L. (eds.) 1975. *European peasants and their markets: essays in agrarian economic history*, Princeton.

Passmore, J.B. 1930. *The English plough*, Oxford.

Pawson, H.C. 1957. *Robert Bakewell: pioneer livestock breeder.*

Peacock, A.J. 1963. *Bread or blood? A study of the agrarian riots in East Anglia in 1816.*

Perkins, J.A. 1976. Harvest technology and labour supply in Lincolnshire and the East Riding of Yorkshire 1750–1850, part one, *Tools and Tillage*, 3: 46–58.

Perkins, J.A. 1977. Harvest technology and labour supply in Lincolnshire and the East Riding of Yorkshire 1750–1850, part two, *Tools and Tillage*, 3: 125–35.

Perkins, W.F. 1939. *British and Irish writers on agriculture*, 3rd edn, Lymington.

Perren, R. 1989. Markets and marketing. In Mingay (1989), 190–274.

Pettit, P.A.J. 1967. *The Royal forests of Northamptonshire: a study in their economy 1558–1714*, Northamptonshire Record Society 23.

Phillips, A.D.M. 1989. *The underdraining of farmland in England during the nineteenth century*, Cambridge.

Phythian-Adams, C. 1993. Local history and societal history, *Local Population Studies*, 51: 30–45.

Pickles, M.F. 1981. Agrarian society and wealth in mid-Wharfedale, 1664–1743, *Yorkshire Archaeological Journal*, 53: 63–78.

Pollard, S. 1965. *The genesis of modern management: a study of the industrial revolution in Great Britain.*

Ponko, V. 1965. N.S.B. Gras and Elizabethan corn policy: a reexamination of the problem, *EcHR*, 17: 24–42.

Pounds, N.J.G. 1973. Barton farming in eighteenth century Cornwall, *Journal of the Royal Institution of Cornwall*, new series, 7: 55–75.

Power, J.P. and Campbell, B.M.S. 1992. Cluster analysis and the classification of medieval demesne-farming systems, *TIBG*, new series 17: 227–45.

Pressnell, L.S. (ed.) 1960. *Studies in the industrial revolution.*

Pretty, J. 1991. Farmers' extension practice and technology adaption: agricultural revolution in 17–19th century Britain, *Agriculture and Human Values*, 8: 132–48.

Prince, H.C. 1964. The origins of pits and depressions in Norfolk, *Geography*, 49: 15–32.

Prince, H.C. 1988. Art and agrarian change, 1710–1815. In Cosgrove and Daniels (1988), 98–118.

Prince, H.C. 1989. The changing rural landscape, 1750–1850. In Mingay (1989), 7–83.

Prothero, R.L. 1888. *Pioneers and progress in English farming.*

Prothero, R.L. 1912. *English farming past and present.*

Public Record Office of Northern Ireland, 1975. *An Anglo-Irish dialogue: a calendar of the correspondence between John Foster and Lord Sheffield 1774–1821,* Belfast.

Purdum, J.J. 1978. The profitability and timing of Parliamentary land enclosure, *EEcH*, 15: 313–26.

Pusey, P. 1839. On the present state of the science of agriculture in England, *Journal of the English Agricultural Society*, 1: 1–21.

Pusey, P. 1850. On the progress of agricultural knowledge during the last eight years, *JRASE*, 11: 381–42

Quinault, R. and Stevenson, J. (eds.) 1974. *Popular protest and public order.*

Rackham, O. 1980. *Ancient woodland: its history, vegetation and uses in England.*

Rackham, O. 1986. *The history of the countryside.*

Ravensdale, J.R. 1974. *Liable to floods: village landscapes on the edge of the fens AD 450–1850,* Cambridge.

Reed, M. 1981. Pre-parliamentary enclosure in the East Midlands, 1550–1750 and its impact upon the landscape, *Landscape History*, 3: 59–68.

Reed, M. 1984a. Enclosure in north Buckinghamshire 1500–1750, *AHR*, 32: 133–44.

Reed, M. 1988. *Buckinghamshire probate inventories 1661–1714,* Buckinghamshire Record Society 24.

Reed, M. (ed.) 1984b. *Discovering past landscapes.*

Reed, M. 1990. 'Gnawing it out': a new look at economic relations in nineteenth-century rural England, *Rural History*, 1: 83–94.

Reed, M. and Wells, R. 1990a. An agenda for modern English rural history?. In Reed and Wells (1990b), 215–23.

Reed, M. and Wells, R. 1990b. *Class, conflict and protest in the English country-side, 1700–1880.*

Richardson, T.L. 1976. The agricultural labourer's standard of living in Kent 1790–1840. In Oddy and Miller (1976), 103–16.

Richardson, T.L. 1991. Agricultural labourers' wages and the cost of living in Essex, 1790–1840: a contribution to the standard of living debate. In Holderness and Turner (1991), 69–90.

Richardson, T.L. 1993. The agricultural labourers' standard of living in Lincolnshire, 1790–1840: social protest and public order, *AHR*, 41: 1–19.

Riddall, J.G. 1988. *Introduction to land law,* 4th edn.

Ritson, C. 1977. *Agricultural economics: principles and policy.*

Roberts, M. 1979. Sickles and scythes: women's work and men's work at harvest time, *History Workshop Journal*, 7: 3–28.

Rogers, J.E.T. 1866–1902. *A history of agriculture and prices in England,* 7 vols., Oxford.

Rose, M.E. 1972. The relief of poverty, 1834–1914, Studies in Economic and Social

History, Basingstoke.

Rosenheim, J.M. 1989. *The Townshends of Raynham: nobility in transition in Restoration and early Hanoverian England*, Middletown, Conn.

Rotberg, R.I. and Rabb, T.K. (eds.) 1985. *Hunger and history*, Cambridge.

Rotberg, R.I. and Rabb, T.K. (eds.) 1986. *Population and economy: from the traditional to the modern world*.

Rowley, T. (ed.) 1981. *The origins of open-field agriculture*.

Russell, E.J. 1913. *The fertility of the soil*, Cambridge.

Russell, N. 1981. Who improved the eighteenth-century Longhorn cow? In Minchinton (1981), 19–40.

Russell, N. 1986. *Like engend'ring like: heredity and animal breeding in early modern England*, Cambridge.

Ryder, M.L. 1964. The history of sheep breeds in Britain, *AHR*, 12: 1–12; 65–82.

Ryder, M.L. 1983. *Sheep and man*.

Salaman, R.N. 1949. *The history and social influence of the potato*, Cambridge.

Samuel, R. (ed.) 1975. *Village life and labour*.

Samuel, R. (ed.) 1981. *People's history and socialist theory*.

Saville, J. 1969. Primitive accumulation and early industrialisation in Britain, *Socialist Register*, 251–2.

Saville, J. 1988. Hammond, John Lawrence Le Breton and Barbara. In Cannon (1988), 177–8.

Scarfe, N. (ed.) 1988. *A Frenchman's year in Suffolk*, Suffolk Records Society 30, Woodbridge.

Schofield, R.S. 1983. The impact of scarcity and plenty on population change in England, 1541–1871, *Journal of Interdisciplinary History*, 14: 265–91. Reprinted in Rotberg and Rabb (1985), 67–94.

Scott Watson, J.A. and Hobbs, M.E. 1937. *Great farmers*.

Searle, C.E. 1986. Custom, class conflict and agrarian capitalism: the Cumbrian customary economy in the eighteenth century, *P&P*, 110: 106–33.

Sharp, B. 1980. *In contempt of all authority: rural artisans and riot in the west of England 1586–1660*, Berkeley.

Sharpe, J.A. 1987. *Early modern England: a social history 1550–1760*.

Sheail, J. 1972. The distribution of taxable population and wealth in England during the early sixteenth century, *TIBG*, 55: 111–26.

Sheppard, J.A. 1957. The medieval meres of Holderness, TIBG, 23: 75–86.

Shiel, R.S. 1991. Improving soil fertility in the pre-fertilizer era. In Campbell and Overton (1991), 51–77.

Short, B. 1984. The decline of living-in servants in the transition to capitalist farming: a critique of the Sussex evidence, *Sussex Archaeological Collections*, 122: 147–64.

Skipp, V.H.T. 1970. Economic and social change in the forest of Arden 1530–1649. In Thirsk (1970b), 84–111.

Skipp, V.H.T. 1978. *Crisis and development: an ecological case study of the Forest of Arden 1570–1674*, Cambridge.

Slack, P. 1988. *Poverty and vagrancy in Tudor and Stuart England*.

Slack, P. 1990. *The English poor law 1531–1782*, Studies in Economic and Social History, Basingstoke.

Slater, G. 1907. *The English peasantry and the enclosure of common fields.*

Slicher van Bath, B.H. 1963. *The agrarian history of western Europe A.D. 500–1850*, trans. O. Ordish.

Smith, A.H. 1989. Labourers in late sixteenth-century England: a case study from north Norfolk, *Continuity and Change*, 4: 11–52; 367–94.

Smith, R.M. (ed.) 1984. *Land, kinship and life-cycle*, Cambridge.

Snell, K.D.M. 1981. Agricultural seasonal unemployment, the standard of living, and women's work in the south and east, 1690–1860, *EcHR*, 34: 407–37.

Snell, K.D.M. 1985. *Annals of the labouring poor: social change and agrarian England, 1660–1900*, Cambridge.

Solar, P.M. 1995. Poor relief and English economic development before the industrial revolution, *EcHR*, 48: 1–22.

Spence, C.C. 1960. *God speed the plow: the coming of steam cultivation to Great Britain.* Urbana, Ill.

Spenceley, G.F.R. 1973. The origins of the English pillow lace industry, *AHR*, 21: 81–93.

Spufford, M. 1974. *Contrasting communities: English villages in the sixteenth and seventeenth centuries*, Cambridge.

Spufford, M. (ed.) 1994. *The world of rural dissenters, 1520–1725*, Cambridge.

Stamp, L.D. 1948. *The land of Britain.*

Steane, J. 1974. *The Northamptonshire landscape.*

Steer, F.W. 1950. *Farm and cottage inventories of mid Essex, 1635–1749*, Essex Record Office Publication 8; revised edn, Chichester, 1965.

Stephens, W.B. 1969. Sources for the history of agriculture in the English village and their treatment, *AH*, 43: 225–38.

Stephens, W.B. 1981. *Sources for English local history*, 2nd edn.

Stevenson, J. 1974. Food riots in England, 1792–1818. In Quinault and Stevenson (1974), 33–71.

Stevenson, J. 1979. *Popular disturbances in England, 1700–1870.*

Stevenson, J. 1985. The 'moral economy' of the English crowd: myth and reality. In Fletcher and Stevenson (1985), 218–38.

Stone, L. 1965. *The crisis of the aristocracy*, Oxford.

Stone, L. and Stone, J.C.F. 1984. *An open elite? England 1540–1880*, Oxford.

Stratton, J.M. 1978. *Agricultural records A.D. 220–1977*, 2nd edn.

Sturgess, R.W. 1966. The agricultural revolution on the English clays, *AHR*, 14: 104–21.

Swales, T.H. 1966. The redistribution of the monastic lands in Norfolk at the Dissolution, *Norfolk Archaeology*, 34: 14–44.

Sykes, J.D. 1981. Agriculture and science. In Mingay (1981), I, 260–72.

Sylvester, D. 1969. *The rural landscape of the Welsh Borderland: a study in historical geography.*

Tarrant, J.R. 1974. *Agricultural geography.*

Tate, W.E. 1967. *The English village and the enclosure movements.*

Tate, W.E. 1978. *A domesday of English enclosure acts and awards.*

Tawney, R.H. 1912. *The agrarian problem in the sixteenth century and after.*

Tawney, R.H. 1941. The rise of the gentry 1558–1640, *Economic History Review*, 11: 1–38.

Taylor, C. 1970. *Dorset.*

Taylor, C. 1975. *Fields in the English landscape.*

Taylor, E.R.G. 1936. Leland's England. In Darby (1936), 330–53.

Terrill, R. 1973. *R.H. Tawney and his times: socialism as fellowship.*

Thirsk, J. 1953. *Fenland farming in the sixteenth century,* University College of Leicester, Department of English Local History, Occasional Paper 3.

Thirsk, J. 1955. The content and sources of English agrarian history after 1500, *AHR,* 3: 66–79. Reprinted in Thirsk (1984a), 1–16.

Thirsk, J. 1957. *English peasant farming: the agrarian history of Lincolnshire from Tudor to recent times.*

Thirsk, J. 1961. Industries in the countryside. In Fisher (1961), 70–88. Reprinted in Thirsk (1984a), 217–34.

Thirsk, J. 1967a. Enclosing and engrossing. In Thirsk (1967d), 200–55.

Thirsk, J. 1967b. Farming techniques. In Thirsk (1967d), 161–99.

Thirsk, J. 1967c. The farming regions of England. In Thirsk (1967d), 1–112.

Thirsk, J. 1970a. Seventeenth-century agriculture and social change. In Thirsk (1970b), 148–77; reprinted in Thirsk (1984a), 183–216.

Thirsk, J. 1973a. The fantastical folly of fashion; the English stocking knitting industry, 1500–1700. In Harte and Ponting (1973), 50–73; reprinted in Thirsk (1984a), 235–59.

Thirsk, J. 1973b. The roots of industrial England. In Baker* and Harley (1973), 93–108.

Thirsk, J. 1974. New crops and their diffusion: tobacco growing in seventeenth century England. In Chalklin and Havinden (1974), 76–103. Reprinted in Thirsk (1984a), 259–86.

Thirsk, J. 1983. Plough and pen: agricultural writers in the seventeenth century, in Aston, *et al.* (1983), 295–318.

Thirsk, J. 1984a. *The rural economy of England: collected essays.*

Thirsk, J. 1985a. Agricultural innovations and their diffusion. In Thirsk (1985b), 533–89

Thirsk, J. 1987. *England's agricultural regions and agrarian history.*

Thirsk, J. 1992. The Crown as projector on its own estates, from Elizabeth I to Charles I. In Hoyle (1992b), 297–352.

Thirsk, J. (ed.) 1967d. *The agrarian history of England and Wales,* IV, *1500–1640,* Cambridge.

Thirsk, J. (ed.) 1970b. *Land, church and people: essays presented to Professor H.P.R. Finberg,* Agricultural History Review, Supplement, 18: 148–77, Reading.

Thirsk, J. (ed.) 1984b. *The agrarian history of England and Wales,* VI, *1640–1750: regional farming systems,* Cambridge.

Thirsk, J. (ed.) 1985b. *The agrarian history of England and Wales,* VII, *1640–1750: agrarian change,* Cambridge.

Thirsk, J. and Cooper, J.P. (eds.) 1972. *Seventeenth-century economic documents,* Oxford.

Thomas, B. 1985a. Escaping from constraints: the industrial revolution in a Malthusian context, *Journal of Interdisciplinary History,* 15: 729–53. Reprinted in Rotberg and Rabb (1986), 169–94.

Thomas, B. 1985b. Food supply in the United Kingdom during the industrial revolution. In Mokyr (1985), 137–50; reprinted as, Britain's food supply, 1760–1846: the Irish contribution, in Thomas (1993), 81–99.

Thomas, B. 1993. *The industrial revolution and the Atlantic economy.*

Thomas, K. 1964. Work and leisure in pre-industrial society, *P&P*, 29: 50–62.

Thomas, K. 1971. *Religion and the decline of magic: studies in popular beliefs in sixteenth- and seventeenth-century England.*

Thompson, E.P. 1971. The moral economy of the English crowd in the eighteenth century, *P&P*, 50: 76–136. Reprinted in Thompson (1991a), 185–258.

Thompson, E.P. 1974. Patrician society, plebeian culture, *Journal of Social History*, 7: 382–405.

Thompson, E.P. 1975. *Whigs and hunters: the origin of the Black Act.*

Thompson, E.P. 1991a. *Customs in common.*

Thompson, E.P. 1991b. The moral economy reviewed. In Thompson (1991a), 259–351.

Thompson, F.M.L. 1963. *English landed society in the nineteenth century.*

Thompson, F.M.L. 1966. The social distribution of landed property in England since the sixteenth century, *EcHR*, 19: 505–17.

Thompson, F.M.L. 1968. The second agricultural revolution, 1815–1880, *EcHR*, 21: 62–77.

Thompson, F.M.L. 1969. Landownership and economic growth in England in the eighteenth century. In Jones and Woolf (1969), 41–60.

Thompson, F.M.L. 1976. Nineteenth-century horse sense, *EcHR*, 29: 60–81.

Thompson, F.M.L. 1990a. Review of Mingay, G.E., *Agrarian history of England and Wales, VI, 1750–1850*, in *EcHR*, 43: 489–93.

Thompson, F.M.L. (ed.) 1990b. *The Cambridge social history of Britain, I, Regions and communities*, Cambridge.

Thorpe, H. 1964. Types of rural settlement. In Watson and Sissons (1964), 358–79.

Thorpe, H. 1965. The lord and the landscape, *Transactions of the Birmingham Archaeological Society*, 80: 38–77. Reprinted in Mills (1973), 31–82.

Thwaites, W. 1985. Dearth and the marketing of agricultural produce: Oxfordshire c. 1750–1800, *AHR*, 33: 119–31.

Thwaites, W. 1991. The corn market and economic change: Oxford in the eighteenth century, *Midland History*, 16: 103–25.

Timmer, C.P. 1969. The turnip, the new husbandry, and the English agricultural revolution, *Quarterly Journal of Economics*, 83: 375–95.

Tivy, J. 1990. *Agricultural ecology*, Harlow.

Toynbee, A. 1884. *Lectures on the industrial revolution in England.*

Trevor-Roper, H.R. 1953. The gentry 1540–1640, *EcHR*, supplement.

Tribe, K. 1978. *Land, labour and economic discourse.*

Tribe, K. 1981. *Genealogies of capitalism.*

Trinder, B. and Cox, J. 1980. *Yeomen and colliers in Telford: probate inventories for Dudley, Lilleshall, Wellington and Wrockwardine, 1660–1750*, Chichester.

Trow-Smith, R. 1951. *English husbandry from the earliest times to the present day.*

Trow-Smith, R. 1957. *A history of British livestock husbandry to 1700.*

Trow-Smith, R. 1959. *A history of British livestock husbandry 1700–1900.*

Tubbs, C.R. 1965. The development of the smallholding and cottage stock-keeping economy of the New Forest, *AHR*, 13: 23–39.

Turner, M.E. 1980. *English Parliamentary enclosure*, Folkestone.

Turner, M.E. 1981a. Arable in England and Wales: estimates from the 1801 crop return, *JHG*, 7: 291–302.

Turner, M.E. 1981b. Cost, finance, and parliamentary enclosure, *EcHR*, 34: 236–48.

Turner, M.E. 1982. Agricultural productivity in England in the eighteenth century: evidence from crop yields, *EcHR*, 35: 489–510.

Turner, M.E. 1984. *Enclosures in Britain 1750–1830*, Studies in Economic and Social History, Basingstoke.

Turner, M.E. 1986. English open fields and enclosures: retardation or productivity improvements, *JEcH*, 41: 669–92.

Turner, M.E. 1989. Parliamentary enclosures: gains and costs. In Digby and Feinstein (1989), 22–34.

Tusser, T. 1984. *Five hundred points of good husbandry*, ed. G. Grigson , Oxford.

Underdown, D. 1979. The chalk and the cheese: contrasts among the English Clubmen, *P&P*, 85: 25–48.

Underdown, D. 1985. *Revel, riot, and rebellion: popular politics and culture in England, 1603–1660*, Oxford.

Valenze, D. 1991. The art of women and the business of men: women's work and the dairy industry c.1740–1840, *P&P*, 130: 142–69.

Venn, J.A. 1933. *The foundations of agricultural economics*, 2nd edn, Cambridge.

Vince, J. 1982. *Old farms: an illustrated guide*.

Wade-Martins, P. 1993. *Black faces: a history of East Anglian sheep breeds*, Norwich.

Wade-Martins, S. 1980. *A great estate at work*, Cambridge.

Wade-Martins, S. and Williamson, T. 1994. Floated water-meadows in Norfolk: a misplaced innovation, *AHR*, 42: 20–37.

Walter, J. 1985. A rising of the people? The Oxfordshire rising of 1596, *P&P*, 107: 90–143.

Walter, J. and Schofield, R. 1989a. Famine, disease and crisis mortality in early modern England. In Walter and Schofield (1989b), 1–74

Walter, J. and Schofield, R. (eds.) 1989b. *Famine, disease and the social order in early modern society*, Cambridge.

Walter, J. and Wrightson, K. 1976. Dearth and the social order in early modern England, *P&P*, 71: 22–42.

Walton, J.R. 1973. *A study in the diffusion of agricultural machinery in the nineteenth century*, Oxford University School of Geography Research Papers, 5.

Walton, J.R. 1979. Mechanisation in agriculture: a study of the adoption process. In Fox and Butlin (1979), 23–42.

Walton, J.R. 1983. The diffusion of improved sheep breeds in eighteenth- and nineteenth-century Oxfordshire, *JHG*, 9: 175–95.

Walton, J.R. 1984. The diffusion of the improved shorthorn breed of cattle in Britain during the eighteenth and nineteenth centuries, *TIBG*, new series, 9: 22–36.

Walton, J.R. 1986. Pedigree and the national cattle herd circa 1750–1950, *AHR*, 34: 149–70.

Walton, J.R. 1990. Agriculture and rural society 1730–1914. In Dodgshon and Butlin (1990), 323–50.

Watson, J.A.S. and More, J.A. 1924. *Agriculture: the science and practice of British farming*.

Watson, J.W. and Sissons, J.B. (eds.) 1964. *The British Isles: a systematic geography*.

Wells, R. 1988. *Wretched faces: famine in wartime England 1793–1801*, Gloucester.

Wells, R. 1990. Social protest, class, conflict and consciousness, in the English countryside 1700–1880. In Reed and Wells (1990b), 121–98.

Wells, R. 1994. E.P. Thompson, *Customs in Common* and moral economy, *Journal of Peasant Studies*, 21: 263–307.

Westerfield, R.B. 1915. *Middlemen in English business particularly between 1660 and 1760*, New Haven, Conn.

Whetter, J. 1974. *Cornwall in the seventeenth century*, Padstow.

Whitlock, R.A. 1965. *A short history of farming in Britain*.

Whitworth, C. 1771. *The political and commercial works of that celebrated writer Charles D'Avenant*.

Wilcox, H.A. 1933. *The woodlands and marshlands of England*, Liverpool.

Wilkes, A.R. 1981. The diffusion of drill husbandry 1731–1850. In Minchinton (1981), 65–94.

Williams, D.E. 1984. Morals, markets and the English crowd in 1766, *P&P*, 104: 56–73

Williams, M. 1970a. *The draining of the Somerset Levels*, Cambridge.

Williams, M. 1970b. The enclosure and reclamation of waste land in England and Wales in the eighteenth and nineteenth centuries, *TIBG*, 51: 58–69.

Williams, M. 1972. The enclosure of waste land in Somerset, 1700–1900, *TIBG*, 57: 99–123.

Wilmot, S. 1990. *The business of improvement: agriculture and scientific culture in Britain, c.1700 – c.1870*, Historical Geography Research Series 24.

Wilson, T. and Andrews, P.W.S. (eds.) 1956. *Oxford studies in the price mechanism*, Oxford.

Wiseman, J. 1986. *A history of the British pig*.

Wolf, E. 1966. *Peasants*, Englewood Cliffs, NJ.

Woodward, D. 1971. Agricultural revolution in England 1500–1900: a survey, *Local Historian*, 9: 323–33.

Woodward, D. 1973. The Anglo-Irish livestock trade in the seventeenth century, *Irish Historical Studies*, 18: 489–523.

Woodward, D. 1977. Cattle droving in the seventeenth century: a Yorkshire example. In Chaloner and Ratcliffe (1977), 35–58.

Woodward, D. 1990. 'An essay on manures': changing attitudes to fertilization in England, 1500–1800. In Chartres and Hey (1990), 251–78.

Woodward, D. (ed.) 1984. *The farming and memorandum books of Henry Best of Elmswell 1642*, Records of Social and Economic History, new series 8, British Academy, Oxford.

Wordie, J.R. 1974. Social change on the Leveson-Gower estates, *EcHR*, 27: 593–609.

Wordie, J.R. 1982. *Estate management in eighteenth-century England: the building of the Leveson-Gower fortune*, Royal Historical Society Studies in History 30.

Wordie, J.R. 1983. The chronology of English enclosure, 1500–1914, *EcHR*, 36:

483–505.

Worlidge, J. 1697. *Systema agriculturae*, 4th edn.

Worlidge, J. 1704. *Dictionarium rusticum and urbanicum*. Reprinted 1970.

Wrightson, K. 1977. Aspects of social differentiation in rural England *c*. 1580–1660, *Journal of Peasant Studies*, 5: 33–47.

Wrightson, K. 1982. *English society 1580–1680*.

Wrightson, K. 1986. The social order of early modern England: three approaches. In Bonfield, Smith and Wrightson (1986), 177–202.

Wrightson, K. and Levine, D. 1979. *Poverty and piety in an English village, Terling 1525–1700*.

Wrigley, E.A. 1985. Urban growth and agricultural change: England and the continent in the early modern period, *Journal of Interdisciplinary History*, 15: 683–728. Reprinted in Rotberg and Rabb (1986), 123–68; and in Wrigley (1987b), 157–98.

Wrigley, E.A. 1986. Men on the land and men in the countryside: employment in agriculture in early nineteenth-century England. In Bonfield, Smith and Wrightson (1986), 295–336.

Wrigley, E.A. 1987a. Early modern agriculture: a new harvest gathered in, *AHR*, 35: 65–71.

Wrigley, E.A. 1987b. *People, cities and wealth: the transformation of traditional society*, Oxford.

Wrigley, E.A. 1987c. Some reflections on corn yields and prices in pre-industrial economies. In Wrigley (1987b), 92–130.

Wrigley, E.A. 1988. *Continuity, chance and change: the character of the industrial revolution in England*, Cambridge.

Wrigley, E.A. 1991. Energy availability and agricultural productivity. In Campbell and Overton (1991), 323–39.

Wrigley, E.A. and Schofield, R.S. 1981. *The population history of England, 1541–1871: a reconstruction*.

Yarranton, A. 1663. *The improvement improved*.

Yelling, J. 1969. The combination and rotation of crops in East Worcestershire, 1540–1640, *AHR*, 17: 24–43. Reprinted in Baker, Hamshere and Langton (1970), 117–37.

Yelling, J.A. 1970. Probate inventories and the geography of livestock farming: a study of East Worcestershire, 1540–1750, *TIBG*, 51: 111–26.

Yelling, J.A. 1973. Changes in crop production in East Worcestershire 1540–1867, *AHR*, 21: 18–34.

Yelling, J.A. 1977. *Common field and enclosure in England 1450–1850*.

Youings, J. 1971. *The dissolution of the monasteries*.

Young, A. 1770. *The farmer's guide in hiring and stocking farms*.

Young, A. 1771. *A course of experimental agriculture*, 2 vols.

Young, A. 1808. *General report on enclosures*.

Zell, M. 1979. Accounts of a sheep and corn farm, 1558–60, *AHR*, 27: 122–8.

Zell, M. 1985. A wood pasture agrarian regime: Wealden agriculture in the sixteenth century, *Southern History*, 7: 69–93.

Zell, M. 1994. *Industry in the countryside: Wealden society in the sixteenth century*, Cambridge.

Index

Cambridge Studies in Historical Geography

*Titles marked with an asterisk * are available in paperback*

258